From Colonies to Country

A HISTORY OF US

BOOK ONE	The First Americans
BOOK TWO	Making Thirteen Colonies
BOOK THREE	From Colonies to Country
BOOK FOUR	The New Nation
BOOK FIVE	Liberty for All?
BOOK SIX	War, Terrible War
BOOK SEVEN	Reconstruction and Reform
BOOK EIGHT	An Age of Extremes
BOOK NINE	War, Peace, and All That Jazz
BOOK TEN	All the People

The picture on the cover tells a lie.
Read this book and you'll learn why.

Oxford University Press

In these books you will find explorers, farmers, cowboys, heroes, villains, inventors, presidents, poets, pirates, artists, slaves, teachers, revolutionaries, priests, musicians—the girls and boys, men and women, who all became Americans....

BOOK THREE

From Colonies to Country

Joy Hakim

Oxford University Press
New York

Oxford University Press
Oxford New York
Athens Auckland Bangkok Bogotá
Bombay Buenos Aires Calcutta Cape Town
Dar es Salaam Delhi Florence Hong Kong Istanbul
Karachi Kuala Lumpur Madras Madrid Melbourne
Mexico City Nairobi Paris Singapore
Taipei Tokyo Toronto Warsaw
and associated companies in
Berlin Ibadan

Designer: Mervyn E. Clay

Produced by American Historical Publications

Published by Oxford University Press, Inc.
198 Madison Avenue, New York, New York 10016

Library of Congress Cataloging-in-Publication Data
Hakim, Joy.
From colonies to country / Joy Hakim.
p. cm.—(A history of US: 3)
Includes bibliographical references and index.
Summary: Covers American history from the French and Indian War to the Constitutional Convention.
ISBN 0-19-507749-0 (lib. ed.)—ISBN 0-19-507765-2 (series, lib. ed.)
ISBN 0-19-507750-4 (trade paperback ed.)—ISBN 0-19-507766-0 (series, trade paperback ed.)
ISBN 0-19-509508-1 (trade hardcover ed.)—ISBN 0-19-509484-0 (series, trade hardcover ed.)
ISBN 0-19-511073-0 (school paperback ed.)—ISBN 0-19-511070-6 (series, school paperback ed.)
1. United States—History—Colonial period, ca. 1600–1775—Juvenile literature. 2. United States—History—Revolution, 1775–1783—Juvenile literature. [1. United States—History—Colonial period, ca. 1600-1775. 2. United States—History—Revolution, 1775–1783.] I. Title. II. Series: Hakim, Joy. History of US; 3.
E178.3.H22 1993 vol. 3
[E188]
973—dc20 93-15818
CIP

9 8
Printed in the United States of America
on acid-free paper

THIS BOOK IS FOR MY GRANDSON SAMUEL DONALD JOHNSON

LAFAYETTE AT VALLEY FORGE

Lafayette to Washington:

Shall I begin by saying some things you know, but may have forgotten? This world you have cut from the wilderness, is a new world, brighter with sun in summer, colder with winter cold than the world I knew. The air's strange-sharp, the voice rings here with a hard ring. I find no man but looks you in the eye and says his thought in your teeth, and means it. This was not known before on this star we inhabit.

—MAXWELL ANDERSON
FROM HIS PLAY *VALLEY FORGE*

THOMAS JEFFERSON SAID:

He who permits himself to tell a lie once, finds it much easier to do it a second and third time, till at length it becomes habitual: he tells lies without attending to it, and truths without the world's believing him.

AND:

When angry, count ten before you speak; if very angry, an hundred.

AND THIS TOO:

When tempted to do anything in secret, ask yourself if you would do it in public; if you would not, be sure it is wrong.

Contents

George Washington

PREFACE

From Colonies to Country

George III was ruler of the American colonies (as well as England), but never even came here for a visit.

England's King George III kept a diary. This is what he wrote in it on July 4, 1776: "Nothing of importance happened today."

If King George had put his ear to the ground that day, he surely would have heard a rumble. The earth must have shaken. Because something very important did happen on July 4, 1776. It happened on the American continent, but it was several weeks before a ship brought the news to England.

This book is about that momentous day and how it came about. To understand that, we need to go back in time, and we need to put ourselves in England's American colonies.

Let's climb into a time-and-space capsule. Come along. We'll cruise over the North American continent. It is early in the 18th century, and this land is much as it has been since the times beyond remembering.

So dense are the trees and grasses over most of the continent that it is hard to see anything else, except for the birds that fill the sky in wide, ribbon-like formations that are miles and miles long. But we can look right through the trees by turning on

Time Travel

Did you know that you can actually see back into time? This is not science fiction; this is real. Here is an explanation. To understand, you need to know that light takes time to travel. Light may seem to go very fast, but in universe time it is not that fast. The light from our planet can take hundreds or thousands of years to reach distant galaxies. So, if you are standing on a planet that is 300 light years away from the planet Earth, and you have a telescope beamed toward earth, you will see our planet as it was 300 years ago. You will be back near the start of the 18th century. (If you want to know more about this subject, check out a book on astrophysics. You will read about shrinking suns called red dwarfs, and magnetic whirlpools called black holes.)

Thomas Cole painted this picture of a landscape in the Catskill Mountains in 1827, nearly a century after the time when our story begins. Yet in this spacious land the tiny figure of the Indian in the foreground seems as solitary and far away from any neighbors as he would have been in 1727—or 1627, or even before.

the capsule's x-ray beam. Do you see those huge bears, those agile mountain cats, the hardworking beavers, and the big-jawed alligators? This land is an animal playground! Look closely and you'll notice tiny clearings. They are people places. Men, women, and children are clustered in small spaces across the land. Most of the people we see are Native Americans; their ancestors have lived here for thousands of years. They have been fortunate, these people who are now being called Indians. They have had the luxury of much land and few people. They have lived in a kind of natural paradise.

That will change. It will be the newcomers who cause the change. Do you see them? A few have made homes in places called California, and New Mexico, and Texas. They have come from Mexico and Spain and have brought horses and other domesticated animals.

But it is the people who are settling a narrow band of land along the Atlantic coast who will cause most of the great changes that are coming. Most of these East Coast newcomers are from England, or from Africa. They have accomplished much in the century since the first of them climbed from their small wooden ships at Jamestown and Plymouth. They have cleared land and planted crops and built towns. They are creating English-speaking colonies where life is freer than it is in the Old World.

But not for the Africans. For them there is no freedom at all. For them there is slavery. Slavery exists around much of the world in this 18th century of ours. Most people say, "There has always been slavery, and there always will be slavery. Let's not worry about it."

These people don't realize it, but they will change their ideas. That July day in 1776 will help cause a huge change in people's minds. Two hundred years from now, people will be shocked at the very idea of slavery. In order to understand what is about to happen, we are going still farther back in time. It is 1730. We are ordinary people, and we don't believe we are as good as our "betters." Yes, we actually call some people "betters." We even dress differently from the rich, fancy folk who call themselves "gentry." They wear ruffles and silk, while we plain folk wear plain clothes. Slaves are just a part—the bottom part—of what is known as the "common herd."

Almost everyone—commoners as well as gentlefolk—seems to think that people are made differently from one another: that some are ordinary and some are extraordinary; that some are elegant and noble, and others are simple and dull. Most of us who live in the 18th century believe that birth determines where you fit in the ladder of humans. Oh, a few people have never accepted that idea—but they are the kind who sometimes end up in jail, or live frustrated lives.

It is that very idea of aristocrat and peasant that will be blasted away on July 4, 1776. Do you see why the earth will shake?

In some European towns, laws actually prevent men and women of the upper classes from being friends with lower-class people. Much of the world is like that. Poor people don't have a chance. These English

No one in the 18th century knows that millions of people are heading for this continent. Those millions will need homes and food, and that will cause big changes in the land itself. One hundred years from now many of the giant trees we see will be gone. Two hundred years from now most of the bears and mountain lions will be in zoos.

These Cherokee leaders and tribesmen visited London in 1730 to sign a treaty with the British and be entertained by George II himself. In the end, the treaty, like so many others, was a way for the British to get more land in America.

In the 18th century America seemed strange, exotic, and alluring to most Europeans. Around 1700 a Flemish artist (from what is now Belgium) drew America as an Indian king living in a tropical jungle. The artist had never been to America, so he had to use his imagination. Artists often use symbols to represent an idea, a person, or a place. Today Uncle Sam is sometimes used as a symbol for the United States.

colonies are like that, too—but only partly. Many people have come here to escape those European ways. In America the son of a Boston candlemaker will become one of the wealthiest and most famous men in the land.

That is what the momentous day in 1776 will be about. It will be about opportunity for all, and about equality, and about fairness. Americans will fight a revolution to make those things possible. But the most important part of the revolution will be "in the minds and hearts of the people."

The revolution will change everything—well, almost everything. No, it won't solve the awful problem of slavery. But it will unleash the idea that will end slavery, and that will bring women's rights, and children's rights, and all kinds of other rights.

The idea is so daring that nothing like it has been heard in governments before. This is it: ordinary people are as worthwhile and valuable and competent as anyone, even as worthwhile as kings and queens. Can you imagine it! No one is better than anyone else. That idea will transform the whole world.

New ideas are already beginning to change us. This vast land has made us independent-minded. It has given us the opportunity to think for ourselves. For most of us plain people, there has never been any place like these English colonies. Can you guess what will happen when King George and the English Parliament decide to get bossy?

We colonists will say, "No way!" and "Goodbye, England!" That's another part of the story of that most important of all days in American history: July 4, 1776. Read on, good friends. We'll be traveling from colonies to country.

An all-American animal, the possum—as the saying goes, up a gum tree.

1 Freedom of the Press

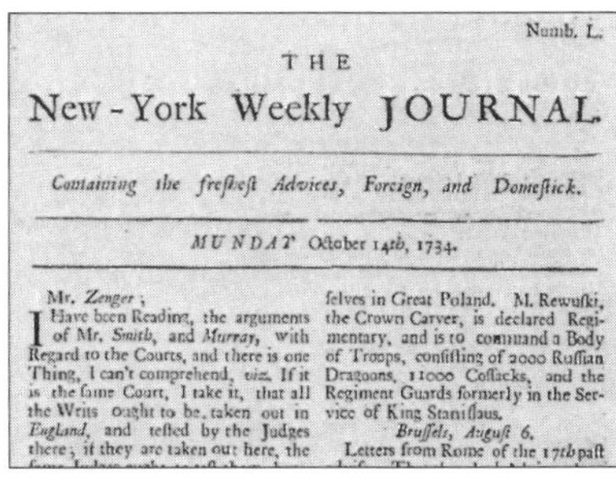

Numb. L.

THE

New-York Weekly JOURNAL.

Containing the freshest Advices, Foreign, and Domestick.

MUNDAY October 14th, 1734.

Mr. *Zenger*;

I Have been Reading, the arguments of Mr. *Smith*, and *Murray*, with Regard to the Courts, and there is one Thing, I can't comprehend, *viz.* If it is the same Court, I take it, that all the Writs ought to be, taken out in *England*, and tested by the Judges there; if they are taken out here, the

selves in Great Poland. M. Rewuski, the Crown Carver, is declared Regimentary, and is to command a Body of Troops, consisting of 2000 Russian Dragoons, 11000 Cossacks, and the Regiment Guards formerly in the Service of King Stanislaus.

Brussels, August 6.

Letters from Rome of the 17th past

The *New York Weekly Journal* was like a news magazine. It printed stories about world happenings as well as events in America.

Peter Zenger was 13 when he sailed to America. He set out from Germany in 1710 with his parents and his brother and sister. They were excited when they left their home; they were looking forward to a good life in a land of freedom and opportunity.

But their voyage was long, and much worse than anyone expected. Some people died on the ship. Peter's father was one of them.

Stop and think how the Zenger children must have felt when the ship finally docked in New York. Their father was dead, they were in a strange land, they had to learn a new language, and they had no money. Their first home was a tent. How would you have felt?

Peter became an apprentice to a printer. He was apprenticed for eight years.

His brother became a carpenter's apprentice. His mother and his sister went to work in a Dutch household helping with the housework. They were indentured servants.

About half the people who came to America in those days became either indentured servants or apprentices. They worked for the person who paid their boat fare—usually from three to 10 years. If they didn't like that person it was too bad. They were stuck for the period of indenture. Some indentured servants were treated just like slaves. An apprentice was usually better off; at least he or she got a chance to learn a trade.

Zenger was apprenticed to William Bradford, a Quaker, who was

One of the writers of the Constitution, Gouverneur Morris (he was not a governor—he just had a strange first name), said, "The trial of Zenger in 1735 was the germ of American freedom, the morning star of that liberty which subsequently revolutionized America."

Peter Zenger's name was actually John Peter Zenger. But, as his brother's name was also John, he was called Peter.

An 18th-century print shop.

one of the best printers in the colonies. The Bradfords treated Peter kindly. During the day he learned the printing trade. At night Peter went to school and Mrs. Bradford helped him with his schoolwork.

When Peter was 21 he was finally free to go out on his own —and he did. First he set up a print shop in Maryland, but later he moved back to New York. There was enough work in that city for two printers.

William Bradford was now publishing a newspaper, the *New York Gazette*. It was the official royal paper; it said only what the English governor wanted said.

Now, I won't go into all the politics of the time—they were just as complicated in the 18th century as they are now—but I will tell you that William Cosby, the king's governor, was a rotten governor. (This Bill Cosby didn't have a sense of humor, and he couldn't stand people who disagreed with him.) Two important lawyers urged Zenger to start a newspaper that would criticize the governor and his politics. They said people should be able to read both sides of a story.

Zenger did it. He founded a newspaper called the *New York Weekly Journal*. It was full of spicy articles. People looked forward to reading it each week. Some articles said Governor Cosby took bribes, took away people's land, and made elections come out the way he wanted them to. The articles were probably written by Zenger's lawyer friends, but no one is sure because they were signed with made-up "pen" names.

Governor Cosby was furious. He wanted the paper to stop publication, but Zenger wouldn't stop. He was arrested and sent to jail. That didn't stop the *New York Weekly Journal*. Peter's wife became the new publisher. The articles criticizing the governor continued.

Peter Zenger stayed in jail for almost 10 months. A trial date was set. Zenger's friends were ready to be his lawyers, but just before the trial, the governor had them disbarred. That meant they could not practice law. Things looked bad for Zenger. The court assigned him a young lawyer with little experience. Then some of Zenger's friends made a secret trip to Philadelphia.

The day of the trial came. It was a hot day in August 1735. The courtroom was full of spectators. Suddenly an old man in a powdered

Pirate Problem

In 1722 the editor of a Boston newspaper, the *New England Courant*, printed an item about pirate ships, hinting that the legislators of Massachusetts might not really be interested in catching them. The article suggested that some of the legislators might be taking money from the pirates. The legislators were furious. The editor was thrown in jail. His name was James Franklin. He stayed in jail for a month. While he was there, his 16-year-old brother ran the newspaper. The brother's name was Benjamin Franklin.

wig entered and stood before the judge. When he said his name there were gasps and whispers throughout the courtroom. He was Andrew Hamilton, a Quaker from Philadelphia, and a friend of Benjamin Franklin and William Penn. Andrew Hamilton was the most famous lawyer in the colonies, and he had come to take Peter Zenger's case.

The attorney general, who was the government's lawyer, said Zenger was guilty of libel. Libel is a crime. You can go to jail for libel.

Today in America, two things are necessary to prove you have committed libel. You must publish something you know is a lie, and that lie must hurt a person. (Let's say that you write an article stating that Mary Smith is a thief. You know that is not true. Mary Smith's boss reads the article and fires her. You have committed libel. She can sue you.) Libel laws were not very clear in the 18th century.

Zenger had printed nasty things about Governor Cosby, but most people thought the things he said were true. "So what?" said the attorney general. It didn't matter if the nasty things were true. Truth was no defense, he said.

Now, in those days it was a crime to say anything bad about the king—even if it was true. The attorney general said the governor was just like the king. He said that gave the governor special rights.

Arbitrary power is power used without considering others.

It is too bad we weren't there to hear what lawyer Hamilton said. They say he spoke very softly, and everyone listened. Some people who were there seemed to know they were hearing history being made. "Free men have a right to complain when hurt," said Andrew Hamilton. "They have a right to oppose arbitrary power by speaking and writing truths... to assert with courage the sense they have of the blessings of liberty, the value they put upon it, and their resolution...to prove it one of the

This is an artist's idea of Zenger's trial. The trial marked the beginning of important differences between libel laws in England and America. People still have more freedom to speak out in the United States than in England.

Andrew Hamilton is said to have drawn the plans for the Philadelphia State House (along with master carpenter Andrew Woolley). The Pennsylvania Assembly met in the building in 1735. Indians in town on business were often housed in its east wing. It was in that handsome State House that a famous declaration was read and approved on July 4, 1776; it changed the world. (Read on to learn more about the Declaration.)

greatest blessings heaven can bestow....*There is no libel if the truth is told,"* he said.

The attorney general wasn't impressed. He kept saying that the jury could only decide facts. They could only decide whether Zenger published the paper. (And of course he had.) The jury was not supposed to interpret the law, or decide if there was libel. The judge would do that, said the attorney general.

"Hold on!" said Andrew Hamilton. Juries are made up of citizens. Citizens are smart enough to decide matters of law as well as matters of fact. And so this jury was. It said Peter Zenger was not guilty.

That case helped give juries the power to decide if the law is being broken. That is an important power, and it will be yours when you are a voting citizen. Every citizen is expected to be ready to serve on a jury. Someday, you will probably be called to be a juror.

But there is more to this case. When Andrew Hamilton spoke to that jury in 1735, he said words that are worth listening to right now: "The question before the court and you, gentlemen of the jury, is not of small nor private concern," said Hamilton. "It is not the cause of one poor printer, nor of New York alone, which you are now trying. No! It may in its consequence affect every freeman that lives under a British government, on the main[land] of America. It is the best cause. It is the cause of liberty."

Today if you want to say that a lawyer is very skilled, you can say he is a "Philadelphia lawyer." That expression dates back to Andrew Hamilton, who came from Philadelphia to fight for Peter Zenger and freedom of the press.

The building at the end of the street topped by the cupola is New York's Federal Hall (no longer standing), where Zenger's trial was held.

2 Jenkins' Ear

George II ruled England when it fought the War of Jenkins' Ear. But he was more interested in money and women than in wars and politics.

By the 18th century Europe had finished with hundreds of years of religious war. It had not, however, finished with war. A new kind of conflict was beginning.

This new kind of war was fought for economic gain. The fight was about land and money, not ideas or religion, and it spread from Europe to America to India to the Philippine Islands. It was the start of worldwide war.

There was something else different about these conflicts: they have been called "people's wars." Before, most wars were started by rulers; often citizens had to be forced to fight them. These new wars had popular leaders and were supported by large numbers of people.

For more than a century, Britain, France, and Spain would fight each other in a series of small wars on several continents. There was King William's War, Queen Anne's War, King George's War, and the French and Indian War. There was also the War of Jenkins' Ear.

I want to tell you about the War of Jenkins' Ear; partly because of that ridiculous name, but also because it was interesting. To begin the story, there was an English sea captain named Robert Jenkins who had two ears—naturally—until the Spaniards got hold of him. Now Jenkins was smuggling slaves onto a Caribbean island where the Spaniards controlled the slave trade. The Spaniards decided to teach Jenkins a lesson. They cut off his ear.

Back in Europe, Spain and England were bickering. Some Englishmen wanted to go to war with Spain. So they brought Jenkins to Parliament and showed what the Spaniards had done to him. That

Talking about money, the English don't have dollars, they have ***pounds.*** Today, a pound is worth about $1.50. A shilling was a small silver coin worth one-twentieth of a pound; at today's exchange rate it would equal about 30 cents. (Don't forget, though, that a shilling bought a lot more in the 18th century than it would now—just as a quarter did in America.) The symbol for a pound, which you may find in this book, is a bit like a script L with a line through it. Here it is: £.

FRENCH
ENGLISH
SPANISH
Hudson Bay
NEW FRANCE
Boston
New York
Atlantic Ocean
LOUISIANA
NEW SPAIN
Santa Fe
San Diego
Pacific Ocean
Charleston
EUROPE CLAIMS AMERICA
FLORIDA
St. Augustine
New Orleans
Gulf of Mexico
INDIAN
ENGLISH
SPANISH
FRENCH

was in 1738; it got a war started. The fighting went on for nine years, with naval battles from North Carolina to Colombia in South America. Since the American colonists were British subjects, many of them fought loyally on England's side.

The war was pretty much of a draw. Yellow fever killed more troops than bullets did. Very little was accomplished except for something that might interest you. For the first time, colonial troops were officially called "Americans" by the English, instead of "provincials" or "colonials." We Americans took pride in that name. When the English said "provincials," they sometimes said it with a sneer. Now they were beginning to respect us.

Captain Lawrence Washington, George's half-brother, was one of those who fought in the war. He served in the West Indies under a popular English admiral, Edward Vernon. After the war, when Lawrence Washington went back to Virginia, he built a plantation house and called it Mount Vernon. But before long, Lawrence died, and that handsome house became George Washington's home. (You can visit Mount Vernon today, just south of Washington, D.C.)

Something else came from that war, something silly. (Sometimes that is all that comes from a war.) Admiral Vernon's nickname was Old Grog, because he wore a coat made of a stiff fabric called grosgrain or grogram. Old Grog made his troops add water to the rum they were drinking because raw rum was making them drunk and disorderly. The sailors began calling that mixture of liquor and water grog. Today we often use the word *grog* to mean any alcoholic liquor.

Well, to sum up, poor Jenkins lost his ear and started a war. A lot of sailors died, and nothing much got decided. And yet everyone knew that Spain, France, and England could not exist together in North America. Spain held Florida, and people in Georgia and the Carolinas didn't much like that. France was powerful in the northern and central regions of the continent. That made the English nervous.

This picture of Admiral Vernon was painted on a pocket watch. It was called a miniature because of its size. Miniatures were usually quick and cheap and were the 18th-century equivalent of a snapshot.

Have you heard the word "groggy"? It means shaky or sleepy. Groggy comes from *grog*, and that word, as you know, comes from Old Grog Vernon. People who drink too much liquor sometimes get the shakes.

It would take a war up north to settle things—at least for a while. That northern war was called the French and Indian War. Some very important things happened during the French and Indian War. Keep reading, and you'll learn about them.

The British navy under Admiral Vernon had a big success at the town of Porto Bello in Panama. They captured the city almost immediately.

3 Frenchmen and Indians

General Braddock was 60 years old in 1755 and had been a soldier since he was 15. But he didn't know how to fight Indians in a wilderness.

In North America the French and Indian War changed the future of the continent. It was a war to answer this question: which would be the strong power in North America—England or France? France, the French colonists, and France's Indian allies fought against England, the English colonists, and England's Indian allies.

The war began with conflicts about land. France and England had real arguments over the same pieces of land. French explorers—Marquette, Joliet, La Salle, and others—had been the first Europeans in the region around the Great Lakes and also in the lands drained by the Ohio and Mississippi rivers. France had sent traders and trappers to those territories, and had set up trading posts as well.

England claimed the same land. In the original English charters, the king granted land from coast to coast—even though no one had any idea where the west coast was. Now that the land along the East Coast was filling up, English-speaking settlers had begun pushing west. Indian hunting grounds were disappearing as the white men moved in. The Indians were alarmed. They were willing to fight to preserve their land.

The English had signed treaties and bought land from many of the Indian nations. But sometimes the treaties were signed without the Indians understanding the details. Indians thought the earth belonged to everyone. One Indian said selling land was like selling the sea, or the sky. And yet, though Indians never owned land individually, Indian tribes did claim the right to use an area of land. It was those rights they signed over to the English.

French traders were at Lake Huron (look at a map) in 1612, eight years before the Pilgrims landed at Plymouth.

The rain, melted snow, and other water that falls on a region flows out of it through streams, rivers, and bigger rivers to the sea. The rivers catching the flow ***drain*** that area.

When George Washington was asked to be a member of Braddock's staff, he replied that he hoped to serve king and country and added, "I wish for nothing more than to attain a small degree of knowledge in the military art."

Why wasn't this war called the French and English War? Can you guess?

It was the English colonists who called it the French and Indian War. They knew who they were; they wanted to remember whom they were fighting. In Europe it is known as the Seven Years' War.

When the English colonists signed treaties with the Indians, the people who signed the treaties usually meant to honor them. The trouble was that the people who actually signed the treaties weren't the ones who lived on the frontier near the Indians. Those frontier people were often rough and rowdy. They wanted land, and sometimes they didn't mind killing for it.

If the Indians had united, perhaps they might have been able to resist the frontier people. But old feuds kept the Indian tribes apart. So when England and France started fighting each other, some Indians sided with the English. Others helped the French. They kept picking at each other—the English, the French, and the Indians—raiding and scalping and killing. Soon the hatred was intense.

New France (Canada) was different from English America, and that made for conflict, too. There was no religious freedom there. The French insisted that all settlers in their territories be Catholic and French. So when 200,000 Huguenots (HUE-guh-notts—though the French said hue-guh-*noes*), who were Protestants, fled from France, many settled in the British colonies. If France had let them settle in Canada, that country would have been stronger. It is easy for us to see that now, but it wasn't so easy then.

France was more interested in the fur trade—and the money it brought—than in settling people on the land. So when English traders began buying furs from the Native Americans and paying high prices for those furs, it made France angry. It hurt their fur business.

Beaver pelts fetched a lot of money in Europe, where they were usually made into hats. French fur traders such as these were ready to fight anyone—Indians or British—trying to take over the trade.

The French were the best friends the Indians had in North America. Mostly they were trappers, traders, and fishermen—like the Native Americans. They understood and respected the land in a way the English never learned. But the Iroquois didn't care. They didn't like them. The Iroquois had been enemies of the French ever since Samuel de Champlain sided against them in their battle with a Huron tribe back in 1609. That was too bad for the French, for the Iroquois led a strong league of six Indian nations. They were the most powerful Indians in Eastern America.

Remember, France and England were both claiming the same territory—especially the lands watered by the Ohio River and its tributaries. The French built forts in that area. One fort was built where Pittsburgh stands today. The French called it Fort Duquesne (dew-CANE). The English said that the fort was in Virginia and the land belonged to them. The governor of Virginia sent a 21-year-old surveyor to tell the French to move on and out.

A surveyor is a person who measures and maps land. This surveyor's name was George Washington. The French told George Washington they were at Fort Duquesne to stay.

Washington and 150 men tried to make them go. They attacked a

General Braddock's army took 32 days to cover 110 miles of forest from Virginia to Fort Duquesne—even with 300 men with axes cutting down trees ahead of them. How many miles a day is that?

Braddock was gloomy about the adventure in America. The night before he left England, he told a young woman friend that he never expected to see her again, and then gave her his will. "We are sent like lambs to the slaughter," he said.

General Braddock died from a shot in the lungs at Fort Duquesne. Two-thirds of his men were killed or wounded.

French scouting party and killed 10 Frenchmen. An English writer, Horace Walpole, said of that small battle, "The volley fired by a young Virginian in the backwoods of America set the world on fire." It was 1754; the French and Indian War had begun.

Washington built a small fort called Fort Necessity. He built it on low ground. When the French attacked, Washington and his men were outnumbered, but they held out until it started raining. Heavy rain flooded the fort, soaked all their gunpowder, and left them defenseless. The French captured the fort, but Washington escaped and learned a lesson he would remember when he became a great general: don't build a camp on low ground.

He learned even more important lessons when he fought with England's famous Major-General Edward Braddock. Braddock arrived in America in 1755. He was expected to push the French out of the Ohio Territory. Braddock decided to begin by capturing Fort Duquesne, and he thought he knew just how to do that. The general had been trained in Europe, on great open battlefields, where armies lined up facing each other and shot long, clumsy guns called muskets. Those European armies seemed awesome. Braddock assumed that European methods would work in America.

George Washington wrote of the British troops in their bright red coats, and the Virginia troops in their handsome blue coats, all marching through the green forest. He said it was one of the most beautiful sights he had ever seen. But he realized those colorful coats were great targets. Braddock didn't. The French and their Indian allies wouldn't fight the kind of war Braddock wanted to fight. They wouldn't stand in a straight line and let the English shoot them. They hid in the woods. They wore skins to camouflage themselves. The Indians screamed blood-chilling war whoops. They shot at the British troops from the woods. The British panicked. They "broke and ran as sheep pursued by dogs," wrote Washington.

The French and Indians were outnumbered almost two to one, but they destroyed the English forces. General Braddock was killed. George Washington escaped with four bullet holes in his coat; two horses were shot from under him. But he learned lessons from Braddock's mistakes.

A young man named Daniel Boone, who drove a wagon in Braddock's army, also noted the way the Indians fought. He had grown up on the frontier, and he could fight and hunt like an Indian. At night Boone sat around the campfire and heard tales of the western lands. He wanted some of that land for himself. He believed the Indians would have to go before he could have it; many others believed as Boone did.

Brown Bess

The most famous gun in the 18th century was the Brown Bess. It fired a flintlock: when you pulled the trigger a small piece of flint snapped against steel. That made sparks, which hit a tiny bit of powder. That lit the powder in a cartridge in the barrel, and that exploding powder sent a round lead bullet flying out of the musket. The Brown Bess wasn't very accurate, but it didn't matter much. Men stood shoulder to shoulder, and it was hard to miss hitting someone.

4 A Most Remarkable Man

Benjamin Franklin told the Albany congress that the American colonies were like the segments of a snake—they had to stick together to survive.

At the very time of the humiliation at Fort Duquesne—when George Washington, Daniel Boone, and General Edward Braddock were defeated by French and Indian foes—a Mohawk Indian was readying himself for a warriors' dance. The Mohawk—named Warraghiyagey (war-rag-ee-YAH-gay)—painted bright designs on his naked chest, stepped into a deerskin kilt adorned with porcupine quills, and donned a cap topped with a single eagle feather. Tied to his wrists and ankles were dried deer's hoofs that rattled as he moved. He ate of ceremonial dog meat and threw a red-painted hatchet onto a war post. Soon he would lead the strenuous dance.

Warraghiyagey means "he who does much."

That Warraghiyagey had hazel eyes and spoke Mohawk with the brogue of an Irishman made no difference to those who would follow him into battle. For they were able to judge a man by what he was, not by what he seemed on the surface. And, clearly, this man was remarkable.

To begin, he was English and loyal to his king. An enormous bear of a man, Warraghiyagey was so full of energy, good spirits, and generosity that even those who thought they would not like him were soon won over. He was both an English American and an Indian American, and he did his best in each of those worlds. No one on this continent has ever done that as well.

William Johnson was a successful businessman as well as an Indian. In the 18th century any man having his portrait painted wore fancy clothes and a wig.

But he was plain William Johnson when he arrived in New

Mohawk sachem Tiyanoga (or Hendrick) became a Christian when he was a young man. He visited England with other Indian leaders in 1710, and was always friendly to the British. The painter of this portrait wanted to show both sides of the man: the Indian leader Tiyanoga, and the British friend Hendrick.

York in 1738, at age 23, from a farm near Dublin. He had no money, but he did have an important relative: an English admiral who sent him up the Hudson River to manage a farm that the admiral had bought from the widow of Governor William Cosby. (Do you remember Governor Bill Cosby?) Johnson soon had land of his own. He became a fur trader and was known to be fair and honest. That was unusual; many white traders tried to cheat the Indians. Johnson's honesty paid off. Before long, he owned more than 30 trading posts, from Detroit to Albany. He has been called America's first chain-store owner. He became rich—immensely rich.

But that is not all. He won two battles that changed American history. After the first of those battles he was knighted by the king. He became Sir William Johnson.

Sir William Warraghiyagey Johnson had a zest for life. He knew how to have a good time. He lived like a feudal lord in a great big mansion, but he never seems to have taken advantage of anyone. In the mid-18th century, William Johnson became one of the colonies' best-known citizens and one of its largest landowners. He was, as I said, remarkable.

But, as you know, when he first moved into New York territory near Albany, William Johnson was an unknown young man with a rich uncle and no money of his own. Right away, he did a very sensible thing. He met his neighbors, the Mohawk Indians, and learned their language. Immediately he liked them, and they liked him. Johnson became a good friend of Tiyanoga, the wise and regal sachem (SAY-chum) who was called Hendrick by the Dutch and the English. Tiyanoga/Hendrick was one of four Indian sachems (the British called them kings) who, back in 1710, went to England and met Queen Anne.

Johnson soon learned the ways of the Mohawk and was named as one of them. Johnson's biographer said, "Sir William was a well-adjusted European man; Warraghiyagey thought and acted as an Indian. These two personalities lived together without strain in one keen mind and passionate heart."

It was a time of jealousy between European peoples. Religious wars had made conditions horrible in parts of Germany, so Germans began moving to New York. The Dutch were already there, and so were the English. Instead of cooperating, they sometimes said nasty

In those days all of Ireland was part of Great Britain, and an Irishman could think of himself as British. That is not so for most of the Irish today.

things about each other—and about the Indians too. Johnson would have none of that. It was the way men and women behaved that was important to him. Anyone kind and decent became his friend. His red-brick manor house always overflowed with people. His wife, Degonwadonti, who was known as Molly Brant, was as bold and intelligent as he. Molly was said to be "handsome" and "uncommonly agreeable" and "a political young lady of the royal blood of the Mohawks." She was known for her skills in forest medicine. Her grandfather was another of the four Indian kings who had been to London to meet the queen. Degonwadonti and Warraghiyagey were married in the Indian way, perhaps in 1757. She was 21, he was 42. It was a happy marriage, and they had seven children.

But I'm getting away from the subject. This chapter is really about the French and Indian War. Remember, the war began badly for the

Turn the book sideways to see where the battle of Lake George took place—on the trail between the Hudson River and the lake, at the very top of the river on the map. The two parts underneath (when the book is right way up) show (left) the English being ambushed by French and Indians, and (right) the French overrun by the determined English and Indian fighters under William Johnson.

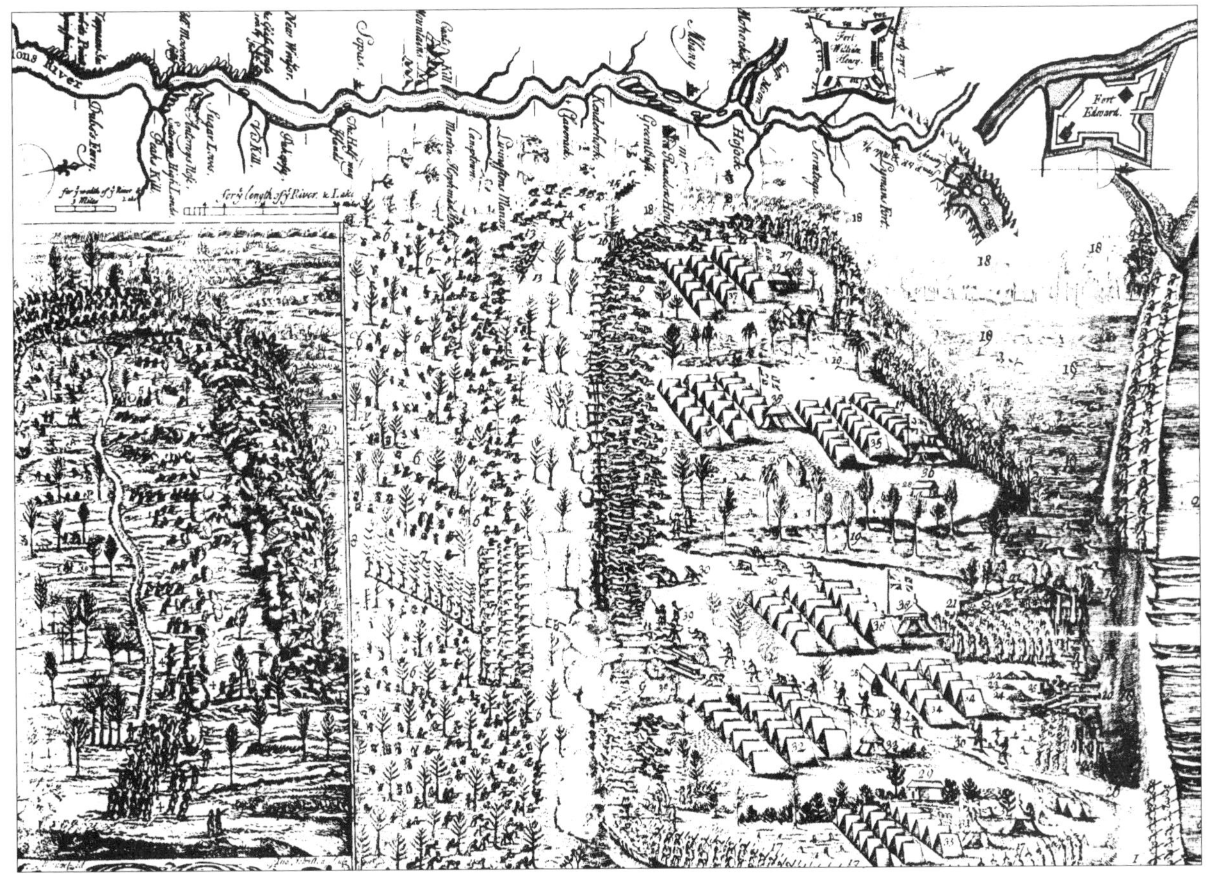

Even though the different groups weren't always as friendly as they should have been, in America they didn't fight and kill over religion as they did in Europe.

English. The French and their Indian allies were better than the English at fighting in the wilderness. Great Britain knew it needed Indian allies if it was going to win this war. In 1754 (which was the same year that George Washington and his scouts killed 10 Frenchmen) some English colonists met in Albany with men from the Iroquois nations. Benjamin Franklin came. William Johnson was there. So was Hendrick, who was much admired by the English.

The reason for the conference was to get the Iroquois as allies. It didn't happen. The Indians wouldn't say yes or no—they just listened. The conference did start the colonial leaders thinking about the Iroquois plan of government. The Iroquois had united six tribes into a confederation. Benjamin Franklin suggested that the colonies unite into a colonial nation. He could see that uniting the tribes had made the Iroquois strong. "It would be a strange thing," said Franklin, "if Six Nations of ignorant savages should be capable of forming a scheme for such a union, and be able to execute it in such a manner as that it has subsisted ages and appears indissoluble; and yet that a like union should be impracticable for ten or a dozen English colonies." In simple English, that means: the Indians have a good system for organizing separate states into a government. Why don't we consider a system like that?

Did you notice that Ben Franklin threw in those words "ignorant savages?" Why would he call them that? Was it because the Indians didn't have written languages and sometimes wore few clothes? Or do you think Ben had his tongue in his cheek (which means he was

Degonwadonti/Molly Brant became a Revolutionary War commander. So did her brother Joseph Brant (left). They both fought for the British (this was after William Johnson had died).

using *irony*—saying one thing when he meant its opposite)? *Ignorant savages* were the words most white people of those days used to describe the Native Americans. But Ben Franklin was never like most people. Besides, he knew many Indian leaders and respected their ideas.

But I'm getting away from the point again. The main reason for the Albany conference was to find a way to solve the problem of French and Indian power. And that hadn't been done.

The delegates to the conference sent a message to the English king: "There is the utmost danger that the whole continent will be subjected to the French."

England had to get the Iroquois to fight on their side. There was only one man who might make that possible: William Johnson. He was named Superintendent of Indian Affairs for the northern colonies.

As Warraghiyagey he called a great meeting. The council fire of the Iroquois League was lit on his property; whole Indian villages came and camped in his yard. The Iroquois were uneasy. They had no wish to fight a white man's war. They had no wish to fight other Indians. Warraghiyagey sat at the council fire. He listened carefully and spoke forcefully; he had learned the Indian art of oratory. Then he did it. He persuaded his Indian friends to fight on the side of the British. He

Even after the French and Indian War was over, there was still bad feeling between the British and many Indians. Here a friendly Abenaki has come to the rescue of an English officer threatened by hostile Indians.

If you look at a map of New York State, near Albany you will see Warrensburg (named for Johnson's uncle, Vice Admiral Peter Warren), and Johnstown (named for William Johnson).

The Indian commissioner for the southern colonies was John Stuart. He had himself fumigated after every meeting with the Indians. He wanted nothing to remain of his contact with "savages."

Baronet is a hereditary title (which means the holder can pass it on to his descendants) granted by the king. A baronet is considered a nobleman. During colonial times, the English named only two Americans baronets.

promised that their land would be protected, and he thought he could honor that promise.

Then Warraghiyagey and his Indian brothers prepared for battle. A French army was on its way to Albany. The French had sailed down Lake Champlain and were now at a lake they called St. Sacrament. Johnson renamed it Lake George in honor of the English king. The French army was led by a German major-general hired by the French because he had won many battles in Europe. His name was Baron Ludwig Dieskau, and he was wily. His well-trained army was composed of French soldiers, Canadians, and Native Americans. General William Johnson had never even seen a battle before. He was on his own with his friend Hendrick, some Indian warriors, and untested soldiers from New Hampshire, Massachusetts, and New York. There were no British soldiers. Hendrick, now an old man, insisted on leading his warriors himself.

Some of the New Englanders, especially those from puritanical Massachusetts, had heard tales of the way Johnson lived: of the Indians who camped on his lawn, and of the grand parties he threw. They disapproved—until they met him. A Massachusetts doctor who fought at Lake George wrote this in a letter home:

> *I must say he is a complete gentleman, and willing to oblige and please all men; familiar and free of access to the lowest sentinel; a gentleman of uncommon smart sense and even temper; never yet saw him in a ruffle, or use any bad language....he is almost universally beloved and esteemed by officers and soldiers...for coolness of head and warmness of heart.*

What happened was astounding. The small army of Native Americans and American colonists beat the French—all by themselves, without the aid of the regular British army. It was a major victory, not easily won. In London people cheered and cheered. And they wept, too, when they heard that the old warrior Hendrick had died in the battle. As did young Colonel Ephraim Williams, who made his will before the fight and left what he owned to start a small college in Massachusetts.

The story of the battle of Lake George was told all over Europe and America. People heard how Johnson was shot in the hip and how he saved the wounded German baron from Indians who wished to scalp him. In Portugal a song was sung of "Wilhelmo Gonson" and his triumph. The painted warrior named Warraghiyagey became a romantic hero, and the English king made him a baronet. That means he was now a knight, and a "Sir."

5 Pitt Steps In

William Pitt, the British foreign secretary, always admired the American colonists' free spirit.

Major-General Jeffrey Amherst (AM-urst) didn't like Sir William Johnson; he didn't like him at all. Amherst was a professional soldier who became commander of England's forces in the northern colonies. He was smart and capable, but also stuffy and haughty. Amherst didn't think much of the American colonists, and he detested Native Americans—he really did believe they were savages. When Johnson was made a baronet, General Amherst was horrified. How could a "provincial"—especially one who ran around in Indian

General Wolfe, the British commander at the battle of Quebec, knew that his best hope lay in surprise. So his troops did something almost impossible—they climbed the Heights of Abraham, the cliffs behind the city, in the dark. In the morning the French were stunned to see the British drawn up outside the city.

New France was surrendered to the English in 1760. It took a few more years for the diplomats to get things settled, but in 1763 the Treaty of Paris officially ended the French and Indian War.

garb—be an English nobleman? But, as I said, Amherst was a capable officer. He knew how to use Warraghiyagey's talents. He knew how to make plans and organize troops.

William Pitt understood that, too. Pitt was foreign secretary (which is the same as secretary of state) in England and one of that nation's greatest statesmen. Pittsburgh is named for him. Pitt intended that this war be won. He sent more English troops to the colonies. Then he looked at a map, and he saw the importance of the St. Lawrence and Niagara rivers. Pitt knew that the French supplied their armies through those two rivers. If the British controlled them, they could keep goods and equipment from reaching the Great Lakes and the Ohio River valley. The French would be like bees cut off from the hive. Pitt told Amherst to take those rivers.

Amherst made plans. He laid siege to a French fort, Louisbourg, which guarded the mouth of the St. Lawrence. That means he would let no one in or out of the fort. After seven weeks of being cut off, the French in the fort were starved. Louisbourg surrendered.

Then Warraghiyagey and the Iroquois won a great battle near the falls of the Niagara River. The English now had control of that river.

Everyone knew that the most important battle would be the one for the city of Quebec. Both sides were confident. Louis Montcalm, the brilliant French general, had smashed the English when he met them before. But he was pitted against England's young general James Wolfe, who was also brilliant. In the middle of the night Wolfe and his English soldiers climbed steep cliffs near Quebec. At sunrise they were on a flat plain behind the city. Montcalm and his men were totally surprised when the English attacked. Both Montcalm and Wolfe were killed in the fighting, but the English won the battle for Quebec.

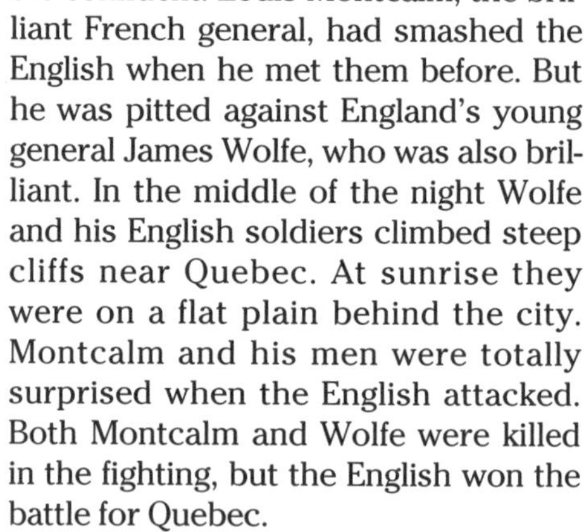

When the city of Montreal fell to a force led by both Sir Jeffrey Amherst and Warraghiyagey, it was all over.

A British soldier came up to General Wolfe as he lay dying at Quebec and said, "They run!" "Who runs?" Wolfe asked. "The enemy, sir!" said the soldier.

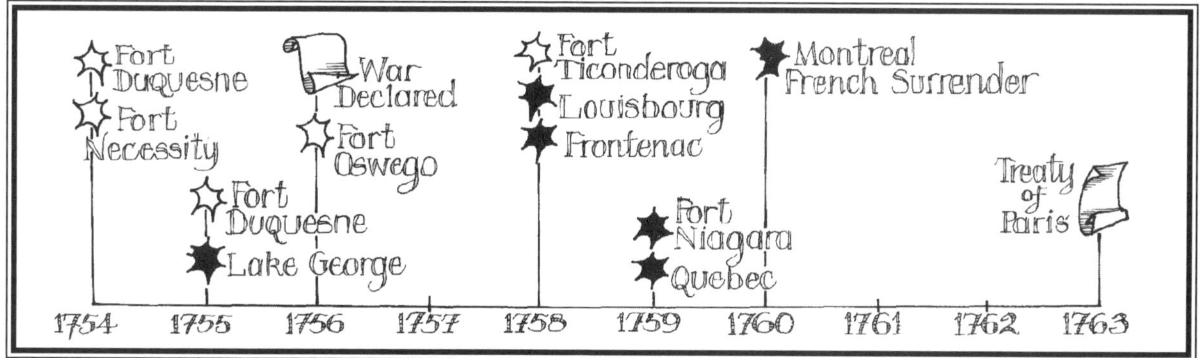

The French and Indian War

The war was very expensive. The British government's spending rose during the war from £6.5 million a year to £14.5 million. Someone had to help pay—and the English thought that the Americans should. That was the beginning of a lot of trouble.

Dates to Stick on Your Timeline

May 1754: Washington leads 150 Virginians to victory over a French exploratory party.

June/July 1754: Albany Congress approves Franklin's Plan of the Union—but not all colonies accept it.

July 1754: Washington surrenders Fort Necessity.

July 1755: Braddock and his English and colonial forces defeated by French and Indians at Fort Duquesne.

September 1755: Johnson and Hendrick triumph at Lake George.

October 1755: 6,000 Acadians leave Nova Scotia because they will not swear loyalty to Britain.

May 1756: the English declare war on the French in Europe, where the war is called the Seven Years' War.

July 1758: General Jeffrey Amherst captures Louisbourg.

July 1759: Johnson's English soldiers and Iroquois Indians capture Fort Niagara.

September 1759: The English win the battle of Quebec. Wolfe and Montcalm are killed.

February 1763: Treaty of Paris ends the French and Indian War.

6 *Au Revoir* (Good-bye), France

Some say that if Sir William Johnson hadn't died in 1774, he might have been commander of the American army instead of Washington.

This is the way the French and Indian War came out: the English won. France was kicked off the North American continent—totally—except for two tiny fishing islands off the coast of Canada.

At the end of the war England claimed all the land from the east coast to the Mississippi. New Orleans and everything west of the Mississippi belonged to Spain. France gave all her western land to Spain in order to keep it out of British hands. That huge territory was called Louisiana. (Spain had helped France for a short time during the war. Later, France got Louisiana back—for a while.)

The British got Florida from Spain, but after a few years Spain took it back. (Does all this sound like a game? It was serious business to those involved.)

The names on the map of North America had changed, but only a few European settlers lived west of the Appalachian Mountains. Most of that

Sir Jeffrey Amherst was commander of the entire British army in the French and Indian War. He lived to be 80 years old and disliked America his whole life long.

Canada Colony

England got a 14th American colony out of the war. The French had called it Acadia; the English named it Nova Scotia. Many French citizens of Acadia were shipped off—against their wishes—by the English to other parts of the colonies. (Henry Wadsworth Longfellow wrote a poem about the Acadians called "Evangeline.") Some Acadians went to what is now Louisiana; they are called "Cajuns." France lost Canada but left her language there. Today many Canadians speak French because their ancestors did. And all Canadian children learn French in school.

western territory was still Indian land, except for some old Spanish colonies across the Mississippi in Texas and New Mexico.

What did the Indians gain from the war? Nothing. For a while there were thanks and treaties and some respect. That didn't last long. Now that France was no longer a threat on the continent, the English colonists didn't need Indians as allies.

For his part in the war, General Amherst was made a knight. He became Sir Jeffrey Amherst. He was also named governor of Virginia, but he wasn't thrilled about that. He said he would accept the post as long as he didn't have to live in Virginia, which was all right with the new king, George III.

The king offered Sir William Johnson the governorship of New York, but he turned it down. His war wounds were acting up, and he wanted to spend time at his mansion, Johnson Hall. Besides, he preferred his job as Superintendent of Indian Affairs in the northern colonies.

Did Sir Jeffrey thank Warraghiyagey and his Indian forces for their help? No. He suggested that the country would be better off if all the Indians were dead. He even came up with the idea of spreading smallpox among them.

Amherst was now the top British military officer in America. He sent out orders to English forts to stop supplying Indians with guns and ammunition and the traditional gifts they used to get from the French. That may sound reasonable, but it wasn't. The Native Americans needed guns in order to hunt. Many had forgotten how to use bows and arrows. They were soon without food or clothing.

The situation got pretty bad. Sir William Johnson begged Amherst to treat the Indians with respect, but Amherst was stubborn. At the same time, settlers were pushing into Indian territory. They were taking tribal land and killing Native Americans. The Indians went on the warpath. Finally people in England got upset. Snobby, mean-spirited Jeffrey Amherst was called back to England in 1763. He was happy to go, because he didn't like Indians or "provincials." English officials tried, as best they could, to keep the settlers out of Indian territory. But, mostly, they couldn't do it.

Indians: Missions Impossible

While the French and Indian War raged in the east, Spaniards were settling in Texas. They were building missions (which were settlements around a church) and presidios (which were forts). Back in 1682 they had built a mission at Ysleta, way over in the western part of today's Texas, near El Paso. They first built at San Antonio in 1718. Soon there were five missions around San Antonio and more in some other places in Texas.

In 1769, a Franciscan priest, Father Junipero Serra, and a group of missionaries built a mission at San Diego in California. Soon there were missions and presidios at Carmel, Monterey, San Francisco, Santa Barbara, and San José, and others in New Mexico and Arizona. The priests intended to convert the natives to Christianity. They didn't intend to kill them—but that was what often happened. The diseases the Spaniards brought wiped them out.

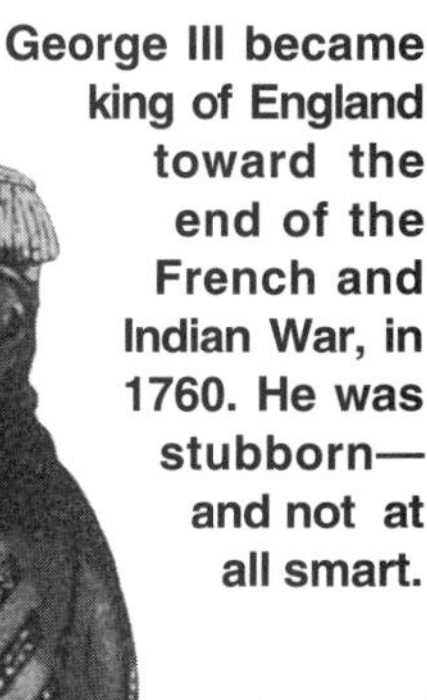

George III became king of England toward the end of the French and Indian War, in 1760. He was stubborn—and not at all smart.

7 Staying in Charge

Dr. Johnson said, "I am willing to love all mankind, except an American."

England now had a big responsibility. She had to manage almost two million people in the 13 colonies, she had to take control of 60,000 French-speaking people in Canada and around the Great Lakes, and she had to keep the English colonists and the Native Americans from killing each other.

The king of England had a great idea for settling the Indian problem. Draw a line right down the crest of the Appalachian Mountains, said the king. Everything to the east of that line would be colonists' territory. Everything to the west would be Indian territory. (Settlers already in the west would have to leave.) And that was what the king ordered in his Proclamation of 1763. If the colonists could be kept east of the Appalachians there would be no more fighting between the settlers and the Native Americans.

That land to the west of the Appalachians wasn't good for much anyway, said most people in England. The learned Dr. Samuel Johnson, who knew a lot about words and wrote the first English-language dictionary, said that the western land that England had gotten as a result of the French and Indian War was "only the barren parts of the continent, the refuse...which the French, who came last, had taken only as better than nothing."

Dr. Johnson should have stuck to his dictionaries. And the king's idea? It sounded wonderful, but it didn't work.

That western land looked mighty good to settlers who wanted farms of their own. The eastern lands were mostly taken. It also looked good to speculators—people who wanted to make money selling land. Benjamin Franklin and George Washington were two of those

Wake Up!

A religious movement, called the Great Awakening, began about 1739, when a spellbinding evangelical English preacher named George Whitefield arrived in America. Thousands of people were converted by Whitefield and by those who followed him. American Protestantism became split between the sedate older sects, the "Old Lights" (Quakers, Anglicans, Congregationalists), and those begun by the new revival preachers, the "New Lights" (Methodists, Presbyterians, Baptists). New Lights reached out to slaves, too; by the end of the century most were Christians.

George Whitefield

who speculated in western lands. Now that the French were gone, the English settlers thought the land should be theirs. A proclamation written in England wasn't going to stop people hungry for land. They kept moving west.

George III said that the lands west of the Proclamation Line (on the map, the row of x's running down the Appalachians) belonged to the Indians. But the English settlers kept moving west and taking the land anyway.

Soon another line was drawn, on the other side of the Appalachians. That was in 1768, in a treaty signed by Sir William Johnson and 14 Iroquois leaders. Johnson hoped to please both sides. The Iroquois got cash and promises; the English-speaking settlers got land over the mountains—especially land west of Albany. But it was just another Indian treaty that would soon be broken. The settlers were on their way west; the Indians who lived west of the Appalachians were doomed to see their way of life destroyed.

Fort Pitt—which had been Fort Duquesne and, before that, the Indian town of Shannopin—became Pittsburgh. At Pittsburgh, two rivers come together and form the mighty Ohio River. From there you can glide to the heart of the continent. It was a gateway to the west.

By 1770 some 5,000 colonists were said to have climbed the mountains to Pittsburgh and then headed on west. They were pioneers, and the first of a river of people who began filling the Ohio River Valley. (Check your map to see where the Ohio River Valley is.) Mostly, these people were

In 1769 Daniel Boone made his first exploring trip to Kentucky. In 1775 he led a group of settlers to that Indian hunting ground. In 1779 Kentucky became a county of Virginia.

There was constant conflict in western Pennsylvania between Indians and settlers. This 1764 cartoon laughed at Benjamin Franklin (left), who tried to help the Indians.

ordinary farm folk who just wanted to make homes for themselves.

Daniel Boone may have helped build the road to Fort Pitt; we know for sure he went to Kentucky:

> *It was on the first of May 1769, that I resigned my domestic happiness, and left my family and peaceable habitation on the Yadkin River, in North Carolina, to wander through the wilderness of America, in quest of the country of Kentucke.*

Thousands followed after Boone cut a path, the Wilderness Road, through the Cumberland Gap. It was a southern route to the other side of the mountains.

Those who went west were a lot like those who had come on the *Mayflower*. They were tough enough to build homes in a strange, raw world. They were able to make their own laws. They were survivors. They were independent-minded. Men and women like that were not likely to take orders from a far-away nation.

8 What Is an American?

"There is room for everybody in America," said Hector St. John Crèvecœur.

Michel Guillaume Jean de Crèvecœur (me-SHELL ghee-OME jahn duh krev-KUR) was a Frenchman with red hair, freckles, a small frame, and a cheerful face. When he was 19 he went to Canada and fought in the French and Indian War on the side of the French. Crèvecœur was a mapmaker for General Montcalm.

But when Crèvecœur saw the English colonies and the freedom they offered, he changed his ideas. He decided to move. He settled on a farm (in what is now New York state) in 1759. Now, you may be wondering about his strange name. It comes from two French words: *crever,* meaning "to break," and *cœur,* meaning "heart." *Crèvecœur* means broken heart. (Remember that.)

When Crèvecœur moved to New York he took an English name: Hector St. John. (He also kept his French name and used it when he went to France. Today we call him Hector St. John Crèvecœur.)

Hector St. John Crèvecœur fell in love with America. He knew that in Europe the aristocrats—wealthy, privileged people—owned most of the land. In America most people were yeoman farmers. That means they owned and worked small farms. Crèvecœur thought farming an ideal life and the English colonies an ideal place—although he also said that some Americans were destroying the land, and that others were always "bawling about liberty without knowing what it is."

Crèvecœur soon married and had a family. He was so happy living on his farm that he decided to write a book about his life in America. In his book he asked a famous question. "What is an American?" he asked.

They Should Have Made Them into Cider

Everyone with an apple tree in his yard—not just farmers like Crèvecœur—made cider, and the Reverend Mr. Whiting of Lynn, Massachusetts, was no exception. He always had a barrel on tap for guests:

And it hath been said that an Indian once coming to his house and Mistress Whiting giving him a drink of cider, he did set down the pot and smacking his lips say that Adam and Eve were rightly damned for eating the apples in the garden of Eden, they should have made them into cider.

American Pie

Crèvecœur didn't just write down beautiful thoughts. He was a farmer and lived off what he grew. You can dry your own apples following his instructions:

You may want to know what we do with so many apples....In the fall we dry great quantities.... Our method is this: we gather the best kind. The neighbouring women are invited to spend the evening at our house. A basket of apples is given to each of them, which they peel, quarter, and core. The quantity I have thus peeled is commonly 20 bushels, which gives me about three of dried ones. Next day a great stage is erected anywhere where cattle can't come. Strong crotches are planted in the ground. Poles are horizontally fixed on these, and boards laid close together. When the scaffold is thus erected, the apples are thinly spread over it. They are soon covered with the bees and wasps and sucking flies of the neighbourhood. This accelerates the drying. Now and then they are turned. At night they are covered with blankets. If it is likely to rain, they are brought into the house. This is repeated until they are perfectly dried.

How would you answer that question?

This is what Crèvecœur said: *The American is a new man, who acts upon new principles; he must therefore entertain new ideas and form new opinions.*

In this new nation, Crèvecœur wrote, people who were mostly the poor and unwanted of other lands forgot Old World hatreds, married each other, and became successful, self-confident citizens. *I could point out to you a man whose grandfather was an Englishman, whose wife was Dutch, whose son married a Frenchwoman, and whose present four sons have now four wives of different nations.*

That kind of thing didn't happen in Europe—especially since most of those people were of different religions.

Crèvecœur was trying to tell the Europeans that something special was happening in this land they had colonized. *We have no princes, for whom we toil, starve, and bleed. We are the most perfect society now existing in the world. Here man is free as he ought to be.*

Crèvecœur said it was opportunity and freedom that had made America *an immense country filled with decent houses, good roads, orchards, meadows, and bridges, where a hundred years ago all was wild, woody, and uncultivated!*

American laws, he said, let people think for themselves. *The law inspects our actions; our thoughts are left to God.* Americans, he said, were good citizens who *carefully read the newspapers...freely blame...governors and others.* (Now, when he talked about blaming governors, Crèvecœur had a certain New York governor in mind. He was Governor William Cosby—you remember him.)

Crèvecœur understood that America had been settled by people who were fed up with Old World problems. Most Americans didn't like societies that kept rich and poor apart. *Here individuals of all nations are melted into a new race of men, whose labors and posterity will one day cause great change in the world.* Crèvecœur was warning Europe. Americans

Crotches are sticks with a fork at one end, so that they will support another stick laid on the fork. A ***scaffold*** is a supporting framework made of poles or pipes. ***Accelerate*** means to speed up.

Dr. Henry Ames of Massachusetts published a popular almanac. In 1762 he wrote:

All men are by nature
equal
But differ greatly in the
sequel.

What did he mean by that?

had new ideas and those ideas might even spread beyond the seas. Would Europe listen to the warning?

Letters of an American Farmer got published in six countries and was immensely popular. It is still read today. But here is something sad: Crèvecœur's name came true. His heart was broken. It happened when Crèvecœur took his eight-year-old son to Europe to get his book published. When he came back to America his wife was dead, his house was burned, and his two younger children were gone. Indians had attacked. He found the children in Boston living with strangers (the Indians had let them go), and they all began a new life. This mended Hector St. John Crèvecœur's heart—at least a bit. But even that broken heart never caused him to lose his faith in the basic decency of all the peoples—Indian, English, French, Spanish, Dutch, African, German, and others—who were forming a new kind of society in the land he loved.

Hector St. John
Crèvecœur introduced alfalfa, a cattle feed crop, to America; he also introduced the American potato to Normandy, in France.

By Crèvecœur's time, the 13 colonies were filling up with large, prosperous farms like this one near Baltimore.

9 A Girl Who Always Did Her Best

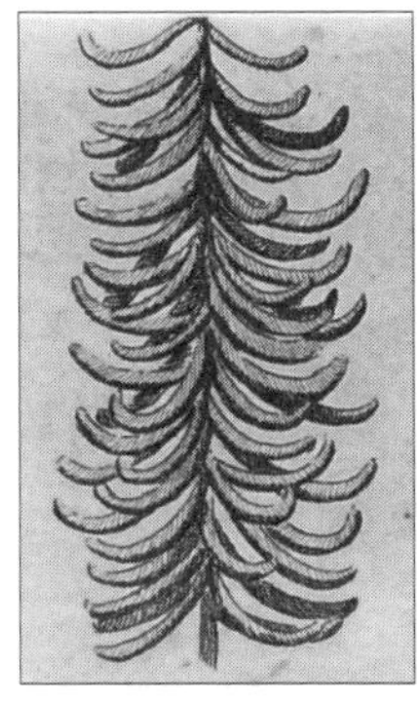

A bunch of indigo seeds. Indigo was used in ancient Egypt as long ago as 1600 B.C.E.

There seemed to be two worlds in the British-American colonies. One was the rough-and-tumble frontier. It was classless—or mostly so. People were judged by what they could do, not by who their parents were. The other world was more orderly. It was a society of rank and class, although it wasn't as rigid as England's class society. Still, there were large landowners (especially in the South), and then merchants, craftspeople, farmers, indentured servants, and—at the very bottom—the slaves. Both of these worlds—the frontier and the structured colonies—produced independent-minded people.

Eliza Lucas was one of those from the privileged plantation world. She was born on the British island of Antigua. (Antigua is in the West Indies, north of Venezuela and east of Puerto Rico.) Her father was a British military officer who believed his daughter should have a good education. He sent her to school in England. She said that good education was the finest gift he could have given her.

When Eliza was 16 her family moved to South Carolina. They had hardly settled down when the War of Jenkins' Ear erupted. Colonel Lucas went off to fight; Mrs. Lucas was ill; Eliza was left in charge. She wrote to a friend in England:

> *I have the business of three plantations to transact...[which] requires much writing and more business and fatigue of other sorts than you can imagine.*

Eliza Lucas was an unusual girl—everyone noticed that. She was a skilled musician, she spoke several languages, and she could think—

An American woman sewed this poem on a sampler in 1736:

Labor for learning
before you grow old
for it is better than
silver or gold.
When silver is gone and
money is spent
then learning is most
excellent.

very well. She had a scientific mind, and she studied and experimented. "I love the vegetable world extremely," she wrote. When she said *vegetable world*, she wasn't talking about peas and carrots for dinner; she meant the whole world of agriculture and plant life.

In South Carolina most planters grew rice, but there was a need for other crops, especially crops that could be sold abroad to bring money into the colonies. Eliza planted fig orchards, and then she dried the figs so they could be shipped far distances. She experimented with ginger, hemp, flax, cotton, alfalfa, and silk. When Eliza's father sent her indigo seeds from the West Indies, she planted them and then replanted the best varieties. Indigo was a valuable blue dye, much desired in Europe. In 1744, Eliza Lucas grew the first successful indigo crop in the colonies. Soon she was giving seeds to other planters. By 1747, South Carolina was exporting 100,000 pounds of indigo a year. That indigo crop, and South Carolina's rice, would become more valuable to England than the gold and silver mines were to Spain. But planting wasn't all Eliza did.

> *I rise at five o'clock in the morning, read till seven, then take a walk in the garden or fields, see that the servants are at their respective business.... The first hour after breakfast is spent at music, the next is constantly employed in recollecting something I have learned, such as French or shorthand.*

In the afternoons she took care of plantation business, wrote letters, and tutored her young sister and two black girls. She was training the girls to teach the other slave children—and that was very unusual in Eliza's day. In her spare time (busy people seem to have more spare time than lazy people) Eliza studied the law and was soon helping others write wills and deal with real-estate contracts.

Eliza Pinckney played an important part in South Carolina's development. But you won't find her in most encyclopedias alongside her sons and nephews. It was hard for women to get into history books.

Scoured the Pewter

It wasn't only plantation owners who worked hard in 18th-century America. In 1775 Abigail Foote of Connecticut wrote up a typical day in her diary:

Fix'd gown for Prude—Mend Mother's Riding hood—Ague in my face—Mother spun short thread—Fix'd two gowns for Welch's girls—Carded tow—spun linen—worked on Cheese Basket—Hatchel'd Flax with Hannah and we did 51 lb apiece—milked the cows—spun linen and did 50 knots—made a broom of Guinea wheat straw—Spun thread to whiten—Went to Mr. Otis's and made them a swinging visit—Israel said I might ride his jade—Set a red Dye—Prude stay'd at home and learned Eve's dream by heart—Had two scholars from Mrs Taylor's—I carded two lb of whole wool—Spun harness twine—Scoured the Pewter.

Ague is toothache. ***Tow*** is short, broken linen thread. Flax (linen) and wool are ***carded*** (combed) to take out tangles. ***Hatcheling*** separates the flax fibers before spinning. A ***jade*** is a horse.

If you read the history of South Carolina, you will get confused. There were three Charles Pinckneys, and one Charles Cotesworth Pinckney. All were leaders. Eliza Lucas married the first Charles Pinckney.

On the Lucas plantations, leather was tanned, barrels made, cloth woven, and shoes sewn. The plantation's own silk was used to make shirts and dresses. Eliza Lucas was overseer of all that activity.

In the meantime, her father was made governor of Antigua. She stayed in charge of the Carolina plantations. Father and daughter wrote loving letters back and forth. But when Eliza's father tried to choose a husband for his daughter (as English fathers often did), she said of his choice, "all the riches of Peru and Chile put together...could not purchase a sufficient esteem for him to make him my husband."

At 22 she chose a husband for herself. He was 45-year-old, dark-eyed Charles Pinckney. Pinckney owned seven plantations and was one of three lawyers who were said to make the decisions that counted in Charles Town. When the royal governor picked a quarrel with the colony's leaders—and called for a new election of legislators—Pinckney and the other leaders just ignored the governor and kept on with their business.

Marrying a wealthy planter didn't mean a lazy life—not for Eliza Lucas Pinckney. She didn't know how to be lazy. She had looms built and began making cloth from flax and hemp. Soon there were two sons and a daughter: Charles Cotesworth, Thomas, and Harriott.

Then tragedy struck. Charles Pinckney died of malaria. Eliza was a widow. Now she had children, the Pinckney lands, and her own lands to manage—all by herself. You won't be surprised to hear that she did an outstanding job with land and family. In addition, she did a lot of reading, thinking, and talking. Some call Eliza Pinckney a Founding Mother. Her sons would play important roles in the trouble that was brewing with England.

Trouble with England? Yes, it would develop, in part because of that independent American spirit. The colonies were producing strong-minded people. Where else could a woman run 10 plantations? But Eliza Pinckney wasn't the only person learning to use her mind on a plantation. It was good training for those who intended to run a nation. George

Washington was running plantations, and so was Thomas Jefferson.

And how about the people who were blazing trails and building homes and clearing farms in a wilderness? And the others who were forming towns, making laws, and starting their own businesses? And the slaves, who were doing much of the hard work of the colonies while developing a culture of song, faith, and survival?

Well, none of these people wanted to be pushed around. If they had to, they would stand up to England. They had seen the English army during the French and Indian War. It wasn't as fearsome as some thought.

Of course, they didn't want to fight England. Most of the colonists were proud and happy to be English. Even those who had come from Scotland, Germany, Holland, and other countries soon thought of themselves as English colonists. They just wanted the same rights that English men and women had in England. That wasn't unreasonable—although the English king, George III, seemed to think it was.

Workers on an indigo farm. Indigo was an ideal second crop to rice in South Carolina, because it needs no work in the winter months, when all the hard work of rice planting must be done.

10 The Rights of Englishmen

King John tried to make the barons pay for his wars. They didn't like that at all.

An ancient story I'll tell you anon,
Of a notable prince, that was called King John
He ruled over England with main and might,
But he did great wrong, and maintained little right.

Just what were those English rights the Americans kept demanding for themselves?

To understand that you need to know there was a time when kings in England could do anything they wanted to do: they could kill people, or take all their land and money, or lock them in dungeons and keep them there.

Some English history will help you understand. Way back in the 13th century, in England, there was a wicked king named John. King John believed he should have total power over everyone. He is said to have arranged for the murder of his own nephew, Arthur, to make sure he would never be king. John even quarreled with the Pope (who was head of the Catholic church and lived in Rome, in Italy). The Pope finally got so angry he closed all the churches in England. That was bad news, especially at a time when the Roman Catholic church was the only Christian church in much of Europe. With the churches closed, no Christian child could be baptized, no one could be legally married, and the dead could not be given a proper burial. Finally the Pope threatened to put another man on the throne of England—popes were powerful enough to do that then—and the king gave in.

But John was a mean sort, and now that he wasn't fighting the Pope he started picking on the English landholders, especially the barons and other noblemen. John, you see, felt that kings had been put on earth by God for men and women to serve.

It was during King John's reign that Robin Hood was said to have lived in England's Sherwood Forest, robbing the rich to help the poor. The rhyme on this page is from an anonymous old English ballad.

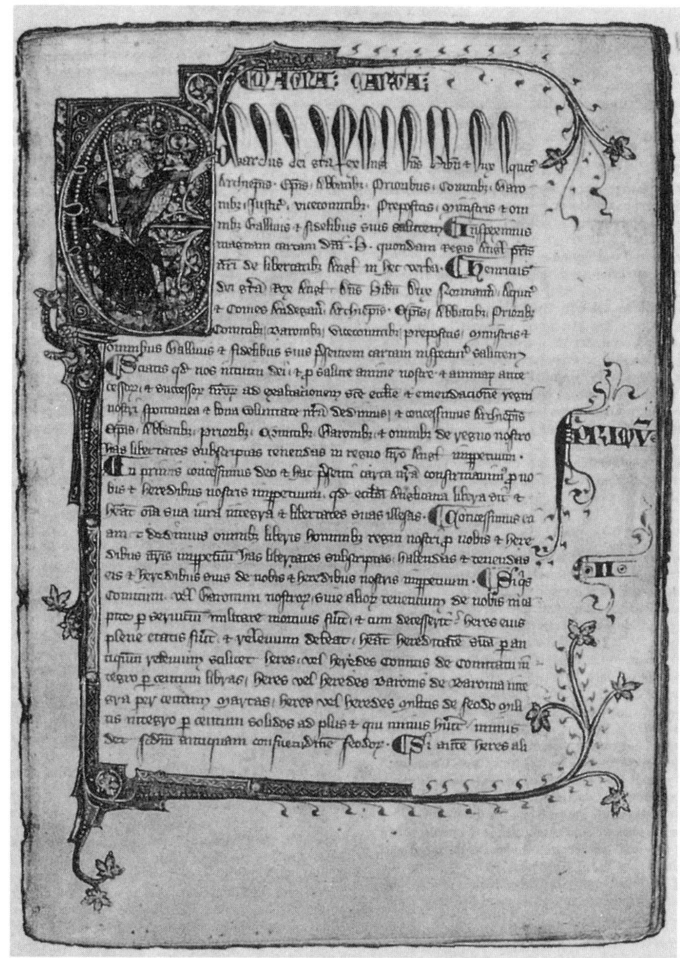

This copy of Magna Carta was handwritten 100 years or so after the barons forced King John to agree to it. John Gutenberg's printing press didn't arrive for another 100 years, in the mid-15th century.

At last the barons could stand no more. In 1215, they captured King John and took him as a prisoner to a tiny island in the Thames (TEMZ) River called Runnymede. There they forced him to sign an agreement that gave Englishmen some basic rights. (No one thought much about women's rights then.)

The agreement said the king could not take land and money from people without Parliament's permission. The agreement also said that no person could be put in jail unless he had a fair trial "by the lawful judgment of his peers, under the law of the land." It granted other rights too. The idea behind that great agreement was that the king's power brought responsibilities. After King John signed his name at Runnymede, kings were no longer free to do anything they wanted.

It was the lords and the wealthy landowners who made the agreement with the king, and they were the ones who, at first, benefited most, but it turned out to be a big step forward for all people. (Before long, men and women were saying that kings were meant to serve the people, not the other way around.)

That document was written in Latin, and its name means great charter; in Latin that is *Magna Carta*. I don't expect you to remember many historic dates, but try to remember 1215. That year is important to people all over the world. Magna Carta is one of the world's greatest documents of freedom. It provided the foundation for many of the rights we enjoy today.

Another very important right the English got for themselves is called "the right of habeas corpus." Now *habeas* and *corpus* are two more Latin words. You might be able to guess that *corpus* means "body." It's similar to our word for a dead body—a corpse. In Latin

Peers are equals, people just like you. Your classmates are your peers. The English barons' peers were other barons.

King William and Queen Mary. She was the niece of Charles II, who gave William Penn the land that became Pennsylvania.

corpus just means a body—dead or alive. And *habeas* means "have," so *habeas corpus* means "have the body." That is what the police must do if they arrest you. They can't lose you (and your body) in a jail. They used to do that. In the old, old days someone could get arrested and thrown in jail, and no one would tell him what he had done wrong. Sometimes the police arrested the wrong person. Sometimes a person got put in jail, and everyone forgot he was there. Sometimes he died in jail without ever knowing why he had been arrested.

If ever you are arrested, the first thing to do is ask for a "writ of habeas corpus." Then you will be brought before a judge, and he will tell you why you are being held. If there is no good reason for your arrest, you can go home. That is a very important right! In many countries today, people still get thrown in jail for no good reason.

Another English right guarantees that your own words can't be used against you in court. Does that sound silly? Why would you say bad things about yourself? Well, if you were tortured you might. You might even say you did something that you didn't do, just to stop the torture. So that, too, is a very important right.

Now, on with history: the English kept adding to their rights and then, in 1688, something revolutionary happened. That something was called the Glorious Revolution, because Parliament got King William and Queen Mary to sign a Bill of Rights that made Parliament more powerful than the king and queen. Since Parliament represented the English people, the people were now more powerful than the monarchs! (Well, maybe not more powerful, but they were headed in that direction. England was on the road to constitutional monarchy.)

All this was especially glorious because no one's head got chopped off during the Glorious Revolution (as had King Charles I's head in 1649).

The English people were very proud of the rights they had won. They had a right to be proud. The American colonists expected those same rights. They were right to insist on them.

The Americans thought of themselves as English citizens living in the colonies. They believed that English rights were their rights. Then things happened that made them think they were losing their precious rights, so they went to war. After the war they wrote a great constitution. It gave the American people basic English rights, and then went

You can read more about the Glorious Revolution in Book 2 of A History of US.

even farther and guaranteed freedoms that no country had ever before given its people.

It took a fight to be free to write that constitution. It didn't have to happen that way, but the British leaders just couldn't see the Americans as equals. They thought they could treat us like little children. But the more they tried to spank the colonists, the tougher we Americans grew.

Magna, the Latin word for *great* or *big*, is the root of some English words connected with greatness or bigness, too—*magnify*, *magnificent*, and *magnate*, for instance. If you do very well in college, you may be awarded a degree *magna cum laude*—which means "with great praise."

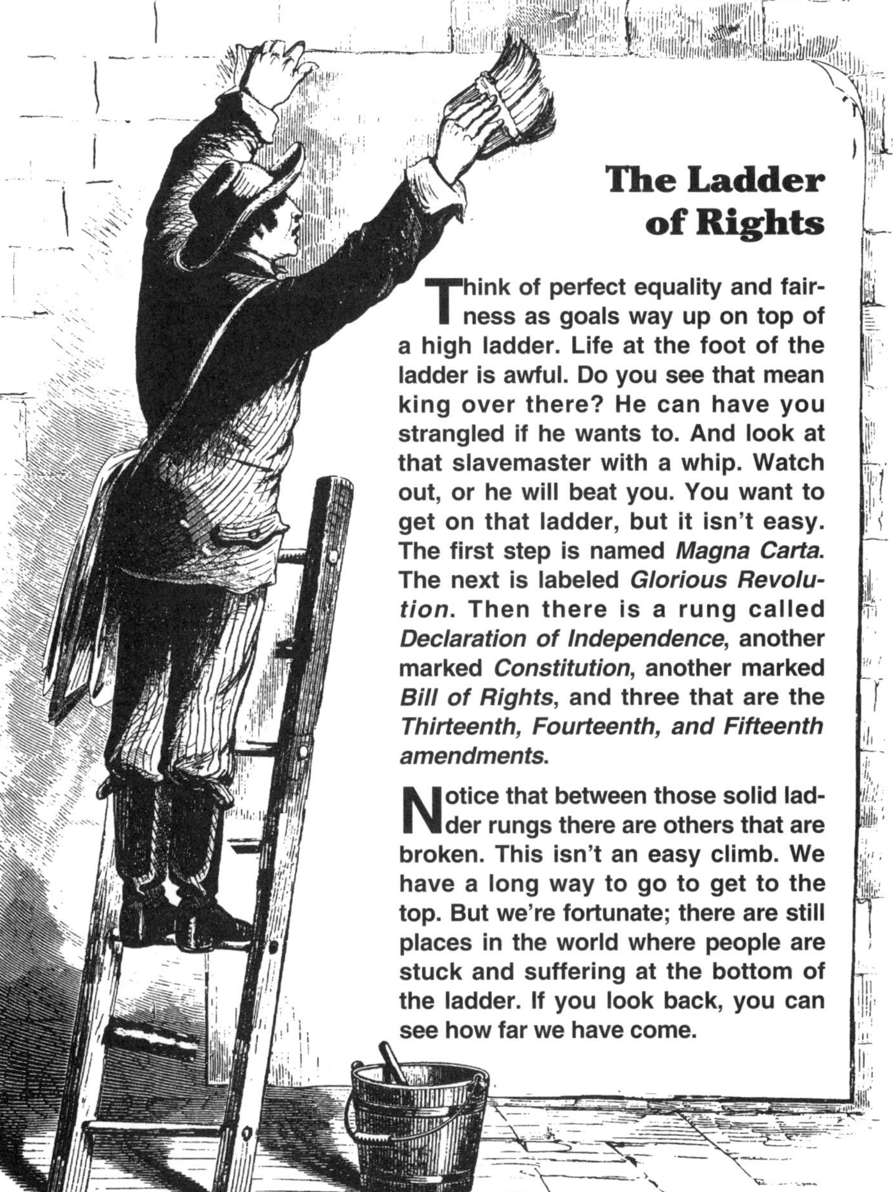

The Ladder of Rights

Think of perfect equality and fairness as goals way up on top of a high ladder. Life at the foot of the ladder is awful. Do you see that mean king over there? He can have you strangled if he wants to. And look at that slavemaster with a whip. Watch out, or he will beat you. You want to get on that ladder, but it isn't easy. The first step is named *Magna Carta*. The next is labeled *Glorious Revolution*. Then there is a rung called *Declaration of Independence*, another marked *Constitution*, another marked *Bill of Rights*, and three that are the *Thirteenth, Fourteenth, and Fifteenth amendments*.

Notice that between those solid ladder rungs there are others that are broken. This isn't an easy climb. We have a long way to go to get to the top. But we're fortunate; there are still places in the world where people are stuck and suffering at the bottom of the ladder. If you look back, you can see how far we have come.

When you read the Constitution, you will see that it was not so magnificent, or great, for the slaves. It did not guarantee their freedom. But it did provide a method for its own improvement. You'll soon learn about that amendment process.

11 A Taxing King

Government officials are sometimes called ministers. They are not ministers of religion. It is confusing, but keep in mind that there are two kinds of ministers: of government and of religion. England's Prime Minister is the leader of the country's government.

Peevish means irritable.

No Right to Tax

William Pitt was an old man, and sick, but he spoke in the House of Commons on January 14, 1766, anyway:

It is my opinion, that this kingdom has no right to lay a tax upon the colonies....The Commons of America [the colonial assemblies] have ever been in possession of...their constitutional rights, of giving and granting their own money....At the same time, this kingdom...has always bound the colonies by her laws, her regulations...in every thing except that of taking their money out of their pockets without their consent. Here I would draw the line.

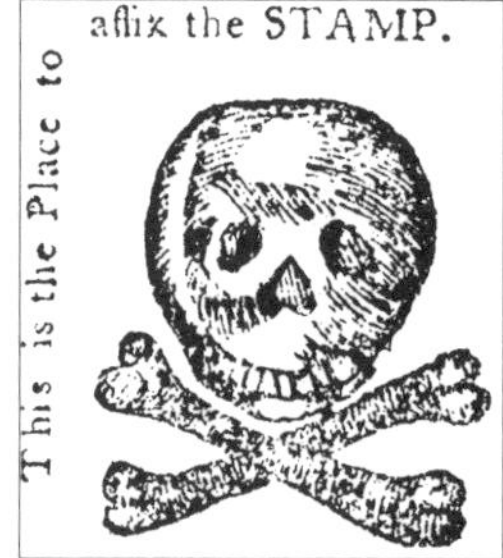

The colonists' opinion of the Stamp Act. On the opposite page, inset, is one of the hated British stamps.

Benjamin Franklin knew that sometimes the best way to get people to think is to make them laugh. So when he was serious, he wrote a joking poem. Here is part of it:

We have an old mother that peevish is grown;
She snubs us like children that scarce walk alone;
She forgets we're grown up and have sense of
our own.

Who was the "old mother?" Why, England, of course! "We" were the colonists. Ben Franklin was right. King George III and his ministers didn't believe the colonists were grown up and capable of ruling themselves. The colonists knew they were. After all, they'd been running most of their own affairs from the time they first arrived in the New World. But even England's William Pitt, who was a friend of America, wrote, "This is the mother country, they are the children; they must obey, and we prescribe."

Part of the problem was that almost none of the English leaders had been to America—or cared to go. They didn't understand the country or its people. One London newspaper called Americans "a mongrel breed."

Now we were, and are, exactly that. A mongrel (MONG-grull) is a dog that is a mixture of breeds—a mutt. From our beginnings, we were a mixture of peoples. That was unusual for a nation. We were attempting something difficult and challenging. The London newspaper thought it was insulting us when it said "mongrel breed." Well, it was no insult—unless, of course, they were calling us dogs.

As I've told you, most colonists (no matter where they came from) thought of themselves as English citizens, so they were hurt by sneers from London. But, to be fair, the colonists didn't quite understand themselves. Even those who had come from England weren't really English anymore. They were now Americans. The people who came to America were different from the stay-at-homes in Europe. Many had risked their lives and gone through great hardships to cross the ocean and build homes and farms in a land of thick forests. They weren't going to let anyone tell them how to run their country. King George never thought about that.

What George and his ministers wanted to do was to teach the colonists a lesson. So they levied taxes, they wouldn't listen when the colonists complained, and they sent soldiers to America. The soldiers had to be housed and fed by the Americans. The British claimed the soldiers were to protect the colonists—but from whom?

At first the Americans were bothered, then they were angered, and then they fought. That fight is called the American Revolution or the War of Independence.

In some ways it really was like a fight between parents and children. Sometimes those kinds of fights come about because parents don't

An English historian said of George III (above), "He was very stupid, really stupid...a clod of a boy whom no one could teach." George was furious when, in 1773, a group of men threw a shipful of tea into Boston harbor rather than pay tax on it (below).

The tea rebels smeared their faces with soot, lampblack, and red ocher.

Party On, George

It was now evening, and I immediately dressed myself in the costume of an Indian, equipped with a small hatchet...with which, and a club, after having painted my face and hands with coal dust in the shop of a blacksmith, I repaired to Griffin's Wharf, where the ships lay that contained the tea....I fell in with many who were dressed, equipped and painted as I was, and who fell in with me and marched in order to the place of our destination....We then were ordered by our commander to open the hatches and take out all the chests of tea and throw them overboard, and we immediately proceeded to execute his orders, first cutting and splitting the chests with our tomahawks, so as to thoroughly expose them to the effects of the water.

—George Hewes, a participant in the Boston Tea Party

realize their children are grown-up and can take care of themselves. Sometimes the children aren't as thoughtful as they could be. There was something to be said for both sides in this quarrel. But, almost everyone agrees, King George made some big mistakes. His pride was more important to him than the valuable American colonies.

George wanted to be a good king. But to be a good king you need some wisdom, and George III didn't have much. He wasn't anywhere as smart as you are.

When George was 10 he was just beginning to learn to read. He never read well. His mother was often heard saying to him, "Be a king, George." Maybe she realized that he wasn't made of kingly material. Later in his life he became ill and hardly able to work. Sometimes he raged and screamed and scared his advisers. That is one of the problems with monarchies: you never know how those royal kids are going to turn out. George III wasn't a bad man; he was actually quite nice. He just wasn't up to the job of being king.

In 1774 a band of angry colonists tarred and feathered the British customs commissioner, Mr. John Malcomb, for doing his job—trying to collect customs duties on goods imported into America.

Remember the Glorious Revolution that gave Parliament more power than the king? Well, George III wasn't happy with that arrangement. He wanted kings to have more power. He chose government officials who seemed to agree with him. One was Charles Townshend (TOWNS-end). Townshend was known to his friends as "Champagne Charlie." He was a likeable man who sometimes got very drunk. Townshend enraged the Americans by sponsoring taxes they thought were unfair.

Nobody likes to pay taxes. In England, people had just protested over a tax on cider. (Cider was a very popular drink. It was alcoholic, unlike the sweet cider you may have tried.) But the British government was having problems with its budget; it needed money. Foreign wars had left England with big bills to pay. The British thought the colonies should help pay some of those bills, especially the ones from the expensive French and Indian War. And maybe we would have, but George III and his ministers didn't explain things well; they just demanded taxes. The colonists knew how European kings and barons taxed the peasants and kept them poor. They didn't want to risk that kind of treatment. People in America began to get nervous and angry.

The colonists kept talking about Magna Carta and English rights. They said Englishmen had the right to vote on their own taxes. They expected that same right. But since no colonists served in Parliament, no colonists got to vote on taxes. The colonists complained that they were being taxed without being represented. They said, "No taxation without representation." That meant they wanted to vote on their own taxes, in their own assemblies, as they had been doing.

King George and his ministers were stubborn. They wanted to show the colonists who was boss. So they levied more taxes.

It was the Stamp Tax (passed in 1765) that enraged most Americans. The colonists were supposed to buy a British stamp for every piece of printed paper they used. That meant every sheet of the newspaper, every document, every playing card—everything. The colonists wouldn't do it. They got so angry they attacked some of the British stamp agents and put tar all over them and then feathers in the tar. It

The actual blend of tea that was thrown overboard at the Boston Tea Party (a mixture of Ceylon and Darjeeling teas, from Sri Lanka and India) can still be bought from the original shippers, Davison Newman of London. The label says, "This tea is from the same London blending House which in the Year of Our Lord 1773 had the Misfortune to suffer a Grievous Wrong in that certain Persons did Place a quantity of its Finest Produce in Boston Harbour."

This famous cartoon portrayed the repeal of the Stamp Act as a solemn funeral for "Miss Stamp." She didn't last long—less than a year.

was a nasty thing to do, but George III and Parliament got the idea. The Stamp Tax could not be collected. It was repealed.

Then "Champagne Charlie" Townshend decided to tax lead, glass, paper, paint, and tea. That upset the colonists so much they decided to get even by not buying anything made in England. It was the English merchants who got angry about that—it cost them a lot of money—and they demanded that the Townshend taxes be repealed. They were, in 1770, except for the tax on tea. It was a small tax, but King George wanted to prove that he and Parliament could tax Americans if they wished to.

To the colonists, that tea tax was an example of taxation without representation. So, in 1773, some people in Boston decided to show King George and Parliament and Lord Townshend what they thought of the tax on tea. They dressed up as Indians and climbed on a ship in Boston harbor and threw a whole load of good English tea into the ocean. Americans called it the Boston Tea Party, but the English didn't. They called it an outrage.

King George was furious! His prime minister, Lord North, ordered Boston harbor closed. Now Lord North was described by another lord as a "great, heavy, booby looking" fellow. He was also called "indolent," which means lazy. Later, a historian wrote that Lord North was "ready to say yes to whatever his royal master might thunder." And poor King George was doing a lot of thundering. The king was ill. Today doctors think he was suffering from a rare disease that affected his mind and emotions: he often lost his temper and went into rages. When he did, Lord North tried to please the king, not the colonists.

Closing Boston harbor meant no ships could enter or leave. That put half the citizens of Boston out of work. They weren't even able to fish in their own waters. Boston lived on its sea trade, and people worried that they might starve.

John Rutledge Visits

What should they do about the Stamp Tax? Twenty-six-year-old John Rutledge sailed from South Carolina to New York in 1765 to meet in a congress with leaders from nine other colonies. Perhaps they could figure out a common plan of action. Young as he was, Rutledge was already known as an eloquent speaker and a power in South Carolina politics.

"It is my first trip to a foreign country," Rutledge wrote to his mother of New York. London was the capital of his world; he had been schooled as a lawyer there.

New York, like his own Charleston, was a great port bounded by two rivers. Each city had about 14,000 residents. But here the differences began. In New York 2,000 were slaves; in Charleston 8,000 were slaves. That fact alone made the cities seem like different worlds.

Rutledge had hardly time to walk about and consider the differences when there was great excitement in the city. Sir William Johnson had arrived with 200 Indians to buy supplies for his 30 trading posts. It was the most important event of the year for New York's merchants. It seemed far

A rioting mob in Boston (left) protests the Stamp Act by throwing the stamped documents on a bonfire.

a Foreign Country

more important—at the time—than a congress called to discuss the Stamp Act. The Iroquois had come down the post road from Albany and a drum and fife corps greeted them as they rode their horses onto Manhattan Island. Johnson camped with his Indian friends in a meadow by a stream near tiny Greenwich Village.

John Rutledge was lodged at the King's Arms tavern, the finest inn in the city. He hired a coach and went to visit Sir William. The Mohawk lord was now one of the most famous of all Americans. Everyone knew he had saved the land west of the Appalachians from French rule.

After they talked of the Stamp Act and politics and England, Sir William told the young Southerner of the Haudenosaunee (ho-dih-no-SHAW-nee), which was the name for the parliament of the Iroquois nations. "If England is to become a great nation," said Johnson, "she must go to school with the Iroquois." By that Johnson meant that the six Iroquois nations had an idea of government that worked so well that the Europeans needed to learn about it. The idea was this: each Iroquois nation governed itself but all were linked together in time of war or when there was business affecting them all. It was that linkage that had made the Iroquois the strongest Indians in the land. Benjamin Franklin had been impressed with that idea when he met with the Iroquois at Albany. Now Rutledge was impressed.

He visited the Mohawk camp again and again, and Sir William came to dinner at the King's Arms. Johnson's ideas startled the brilliant young lawyer. In South Carolina no one thought of Indians as partners on the continent. Rutledge listened and learned.

New York no longer seemed a foreign nation. He had more in common with these colonists from places like Boston and Philadelphia and Albany than he had imagined possible. The delegates to the Stamp Act Congress sent a petition to England's Parliament saying that colonial taxes should be raised only by colonial legislatures. That was the way it had always been done. Besides, it was a right of all British subjects to have their own representatives voting on their own taxes. John Rutledge approved the petition.

Then he went home, taking some new thoughts to South Carolina. Soon he would head north again, to Philadelphia, for a much bigger congress.

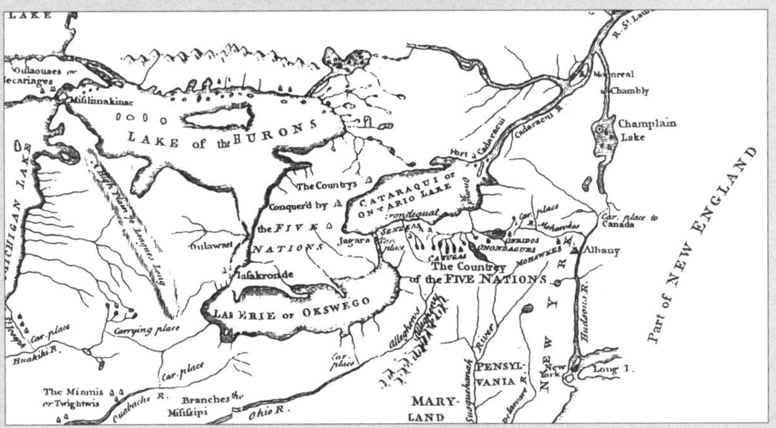

The country of the five Iroquois nations—before the sixth, the Tuscaroras, joined.

Suddenly the other colonies, which had never before paid much attention to one another, all felt sorry for Boston and angry with the king and Lord North. They sent supplies and encouragement. Connecticut sent money; South Carolina sent rice; New York sent sheep. Virginia set aside a day to pray for Boston. During that day the Virginians began to talk about independence. At first the colonists had just wanted England to treat them like grownups. Now that wasn't enough. Now they were thinking seriously about breaking away, about being free.

In 1773, Parliament granted the East India Company exclusive rights to sell tea in the colonies —for less than the tea the Americans smuggled in. That was what really angered Boston's merchants.

12 The Firebrands

After the Boston Massacre (you'll read about it in the next chapter), Samuel Adams made an official protest to the royal governor of Massachusetts.

Can you guess what a firebrand is? Firebrands were very useful when people didn't have matches and the only heat in a house came from the fireplace.

You have probably figured it out. A firebrand is a stick of wood with a spark of fire at one end.

Now, if you look in the dictionary you will see another meaning for firebrand. A firebrand can be a hothead: someone who sparks a revolution, someone who lights a fire in people's minds and hearts.

Historians say the American Revolution had three firebrands: Samuel Adams, Patrick Henry, and Thomas Paine. That war of independence might have happened without them, but it certainly would have been different.

Patrick Henry was a great speaker; Tom Paine was a great writer. Samuel Adams could write well and think well, but what Adams really was was a super busybody. He got everyone keyed up, inspired, and moving.

Sam Adams was a New Englander from Boston with a Puritan background. Tom Paine came from England and lived in the Middle Colonies, in Philadelphia. He was a deist with Quaker friends. Patrick Henry was a Southerner, an Anglican, a Virginian, and a country boy. These men were very different from one another, but alike in one important way: each understood, before most other Americans did, that a break from England was necessary.

A ***deist*** (DEE-ist) believes in God as creator of the universe. Deists reject the idea of an active God who directs worldly events.

In 1770 the annual meeting of New England Quakers prohibited slave-owning—the first American organization to do so.

I've told you before that it took a long time for the colonists to think of themselves as Americans. They thought of themselves as English colonists. Even those who came from France or Germany or Holland soon thought of English rights as their rights.

When they stopped seeing themselves as English, they began to say they were Virginians, or New Englanders, or Carolinians. It was hard for them to understand that they could all be part of the same country. To begin with, they didn't know each other. That was because overland travel between colonies was very difficult. There were no good roads and few bridges. On the fastest stagecoach you could make eight miles an hour—as long as there were no ruts in the road, or mud, or ice. For poor people, travel meant going on foot. But if you were like most travelers, you rode horseback. If you needed to cross a river you usually had to find a boat. Your horses had to swim the river. If you had a lot of baggage, it might take many trips to get it all across. If the river current was swift, you could lose everything—even your life.

By 1760, with good winds and good luck, you could sail from Baltimore to London in four weeks. So wealthy Marylanders were more likely to go to England than to Massachusetts. And the same was true of the Virginians and the South Carolinians. London still seemed the most exciting city in their world. Now can you see why most people in the different colonies were strangers to each other?

Well, the firebrands helped change that. Sam Adams started something called "committees of correspondence." They were groups of prominent citizens who wrote back and forth between colonies and helped each other with problems. They began to be friends.

Adams started other groups, such as the Sons of Liberty. In Boston, the Sons met under an old elm tree that Adams called the Liberty Tree. As soon as the British got a chance they chopped that tree down. (A liberty tree still stands in Annapolis, Maryland.)

But mostly what Sam Adams started was trouble for the British. He was a rabble-rouser and an agitator—a real firebrand—who helped brew up the Boston Tea Party and the fight against the Stamp Act. But you never would have known that to look at him.

Sam Adams was a Humpty Dumpty–looking man: rumpled and pudgy. He came from a prominent Boston family, but he lost almost all the family money because he didn't care about business. Some people said that he was lazy, but he wasn't lazy when it came to fighting for freedom.

The English called Adams a public enemy, an outlaw, and a rebel. They wanted to hang him. He certainly was a troublemaker, but Sam

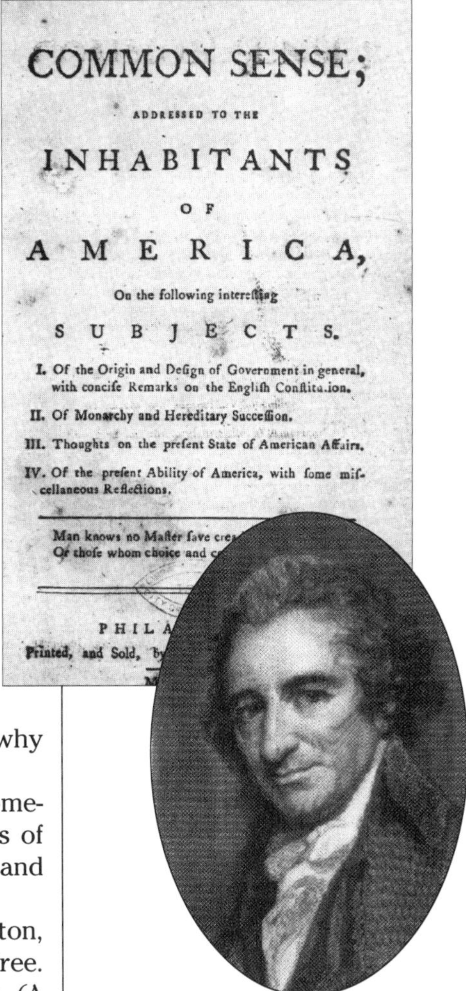

COMMON SENSE;

ADDRESSED TO THE

INHABITANTS

OF

AMERICA,

On the following interefting

SUBJECTS.

I. Of the Origin and Defign of Government in general, with concife Remarks on the Englifh Conftitution.

II. Of Monarchy and Hereditary Succeffion.

III. Thoughts on the prefent State of American Affairs.

IV. Of the prefent Ability of America, with fome mifcellaneous Reflections.

Printed, and Sold,

On the title page of Tom Paine's *Common Sense* runs a couplet: "Man knows no Master save creating Heaven, / Or those whom choice and common good ordain." What does that mean?

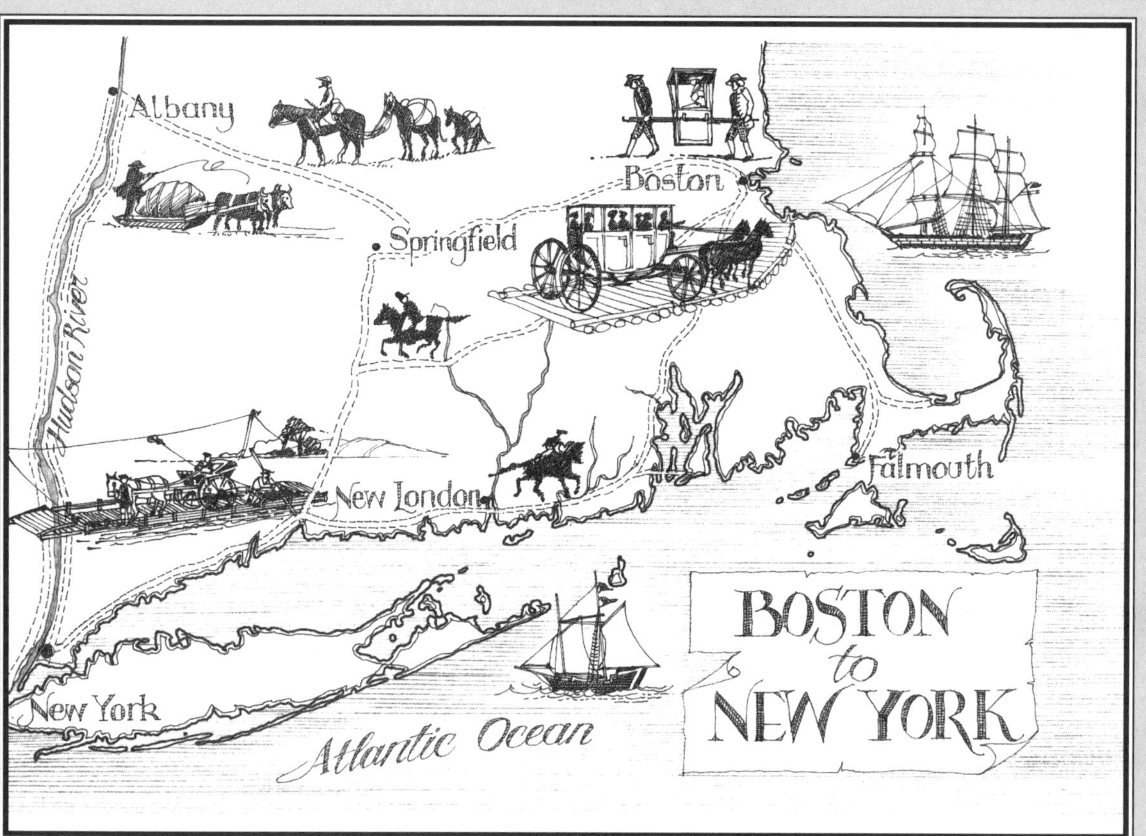

Albany

Boston

Springfield

Hudson River

New London

Falmouth

New York

Atlantic Ocean

BOSTON to NEW YORK

On the Road, or Getting Around in Colonial Times

Until the middle of the 19th century, most American roads were made of dirt. Some were surfaced with gravel or oyster shells. With ice or snow on them they often became impassable. Spring thaws made them turn to mud. Even when they were dry and hard they were full of holes. A horseback rider could make his way, but for a heavy wagon or coach it was a disaster—wheels and axles broke or got mired in the mud.

A post rider blows a horn to announce his arrival in a town.

When Benjamin Franklin became postmaster of the colonies he improved the roads so the mail could be carried more efficiently. These new roads were called post roads.

Some roads were made with rough logs; dirt was put on top of the logs. These roads were called *corduroy* roads. They were so rough that sometimes they made horses go lame and jolted wheeled vehicles into pieces.

There were good roads in ancient America. They were built by the Incas. The Inca roads were 25 feet wide and made of stone and asphalt with retaining walls, suspension bridges, and a series of watchtowers. The roads stretched thousands of miles, over mountaintops and across ravines, from Ecuador to Peru. They were constructed before the arrival of Christopher Columbus.

Adams was different from other rebels in other times. He wanted more than just separation from England. He was inspired by a grander idea: the idea that America could be a special nation where people would be free of kings and princes. A nation where, for the first time in all of history, people would truly rule themselves.

His Puritan ancestors had described their colony as an experiment. They had hoped it would be a close to perfect society. They called it "a city on a hill," and they meant that others should see it and that it would be an example to the whole world. But the Puritan dream was only for Puritans. Sam Adams had a great dream that was for all people.

So did Tom Paine. Now, Sam Adams was a Harvard man who had old roots in the young country. Tom Paine was hardly off the boat from England, in 1774, when he became a firebrand of revolution. He didn't plan it that way. "I thought it very hard," he wrote, "to have the country set on fire...almost the moment I got into it."

He had been apprenticed to a corsetmaker when he was a boy in England. Corsets are tight undergarments that women wore to hold in their stomachs and make their waists look tiny. Being a corsetmaker was not exactly an exciting career—not for a boy with a mind like Tom's. So he ran away and went to sea. But that didn't work out. Then he tried to be a grocer, and a schoolteacher, and a tobacco seller. Nothing worked—except his fine mind, and that kept him learning. He went to lectures and he read everything he could find to read. When he met Ben Franklin in London, he knew he wanted to go to America.

He almost didn't make it. He caught a fever on board ship and was carried ashore half dead, but he was young and strong and he survived. Franklin had given him a letter that got him a job as a writer and magazine editor in Philadelphia. That was the perfect use for his talents, for he was a magician with words.

By the 1770s, the colonists were beginning to want to separate from England. But they weren't quite sure why, and they wondered if it was right to do so. Tom Paine was able to say clearly what people really knew in their hearts. He wrote a pamphlet called *Common Sense*. In it he told the colonists three important things:

- *Monarchy was a poor form of government and they would be better off without it.*
- *Great Britain was hurting their economy with taxes and trade restrictions.*
- *It was foolish for a small island 3,000 miles away to try to rule a whole continent.*

A wagon negotiates a corduroy road. These were laid when a path was too muddy or steep to be traveled any other way. But it was like riding on a giant washboard.

The first stagecoach from New York to Boston began operation in 1772. The trip took six days. Travelers slept in their clothes at inns along the way. They could expect to be waked at three in the morning and to spend 18 hours each day traveling.

A ***monarchy*** is a government headed by a king or queen.

If you send a package to someone, you will pay *shipping* charges. We talk of *shipments* of freight. That's because in early America most freight was sent by oceangoing ships or riverboats. The 13 colonies hugged the Atlantic Ocean.

This is how an artist imagined Patrick Henry's 1765 speech against the Stamp Act to Virginia's House of Burgesses. On the floor in front he has thrown down a gauntlet—a glove—which is a traditional challenge to fight.

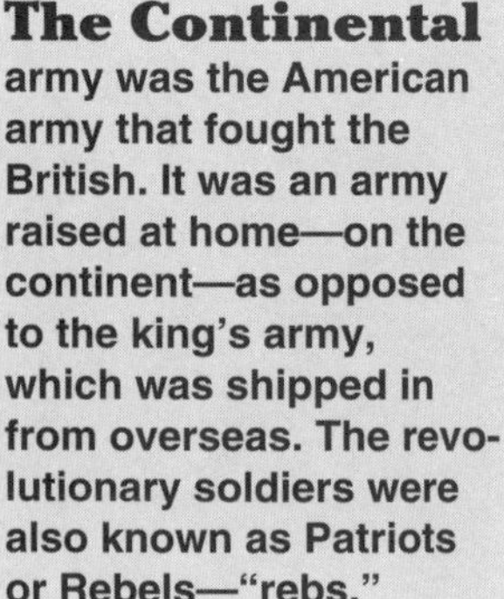

The Continental army was the American army that fought the British. It was an army raised at home—on the continent—as opposed to the king's army, which was shipped in from overseas. The revolutionary soldiers were also known as Patriots or Rebels—"rebs."

Well, of course, all that made common sense. But Tom Paine said it so eloquently that a whole lot of copies of *Common Sense* were sold in a very short time.

When the revolution began, Paine enlisted in the Continental army. Then he wrote a series of pamphlets about the war. He started one of them with these words: "These are the times that try men's souls." Stop and read that line again. "These are the times that try men's souls." What do you think of that as a way to start a book about war? Tom Paine made people stop and think. He was a man of deep beliefs. He believed in the American cause. He was not rich, but he gave a third of his salary to help Washington's army, and he never took any money for his patriotic writings. He said that would demean them.

To ***demean*** means to lower the worth of something.

"We have it in our power to begin the world again," wrote Paine, and he really meant it.

Patrick Henry was the third firebrand, and, like Adams and Paine, he was a failure at first. He was born on a Virginia frontier farm. His father came from Scotland and had been to college there. He taught his son to read—both English and Latin. They read the Bible aloud together, and Patrick learned to love the sounds of the English language. He used English as no American speaker had done before him. He was called a "forest-born Demosthenes," and that was a compliment, because Demosthenes (dih-MOSS-thin-eez) was a great orator and a fighter for freedom in ancient Greece.

But Patrick Henry started out as a storekeeper, and then tried being a planter, and failed at both. Perhaps it was because he had a "passion for fiddling, dancing, and pleasantry." Finally, he studied law and spoke so well as a lawyer that he soon entered politics. He was elected to the House of Burgesses, which met in Virginia's capital, Williamsburg. And that was where he was when the Stamp Act was passed. He was young, but he stood up and said what he thought—that the stamp tax was a threat to liberty. Some of the older Virginians cried, "Treason!" because he was attacking the king. To that Patrick Henry is supposed to have answered, "If this be treason, make the most of it."

Some townsfolk and students from the College of William and Mary stood in the doorway of the House of Burgesses and heard that speech. Among them was a young lawyer named Thomas Jefferson. He never forgot it. He said Henry spoke "as Homer wrote," and Jefferson thought Homer the greatest of writers.

When the English governor of Virginia heard about Patrick Henry's speech, he was furious. He dissolved the House of Burgesses. (That means he told the members to go home.) But that just made them angry. They walked over to the Raleigh Tavern, where they kept on meeting. And Patrick Henry kept talking.

By 1775, 10 years after Henry's Stamp Act speech, it was no longer safe for the burgesses even to gather in Williamsburg. So they met in a church in Richmond. It was there that Patrick Henry gave his most

You can imagine how George III felt when he read Tom Paine's books.

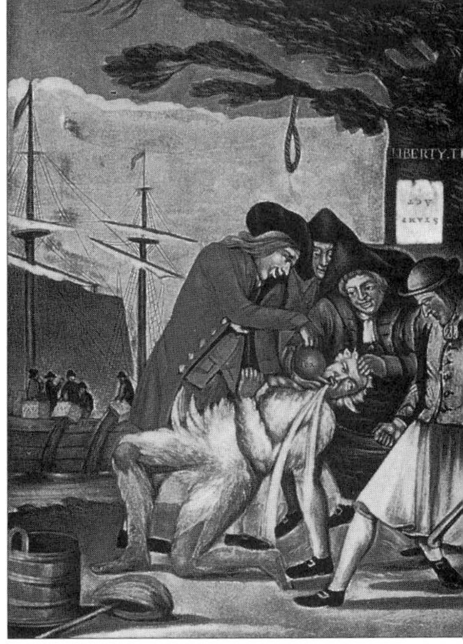

A noose hangs ominously from the Liberty Tree as a mob of colonists makes a royal official drink a pot of tea.

If you've read any books of Greek myths, legends, or history, you've probably read some of the stories that Homer told. Homer lived in ancient Greece and wrote two stories, the *Iliad* and the *Odyssey*, which are still among the most exciting tales of heroism and adventure ever written.

Someone who heard Patrick Henry said, "The tendons on his neck stood out white and rigid like whipcords. His voice rose louder and louder, until the walls of the building seemed to shake....Men leaned forward in their seats...their faces pale."

famous speech. The port of Boston was closed, English soldiers were in the city, and the Massachusetts Assembly had been dissolved. Would Virginia sit idly by?

Henry stepped into the aisle, bowed his head, and held out his arms. He pretended his arms were chained as he began calmly, "Our chains are forged, their clanking may be heard on the plains of Boston." His voice strengthened as he said, "The war is actually begun. The next gale that sweeps from the north will bring to our ears the clash of resounding arms. Our brethren are already in the field. Why stand we here idle?...Is life so dear, or peace so sweet, as to be purchased at the price of chains and slavery?"

Then Patrick Henry threw off the imaginary chains, stood up straight, and cried out, "Forbid it, Almighty God! I know not what course others may take, but as for me, *give me liberty, or give me death!*"

Fellow Fiddlers

Thomas Jefferson first met Patrick Henry at a fiddling session. No, they weren't fiddling around; well, maybe they were.

Gangly, carrot-headed Thomas Jefferson was on his way to the College of William and Mary in Williamsburg. Sturdy Patrick Henry, who was 23, was a not-very-successful shopkeeper.

"On my way to the college, I passed the Christmas holydays at Col. Dandridge's in Hanover," wrote Jefferson. (Colonel Nathaniel Dandridge was a close relative of Martha Washington.) "During the festivity of the season, I met him [Patrick Henry] every day, and we became well acquainted, although I was much his junior, being then in my seventeenth year, and he a married man. His manners had something of a coarseness in them; his passion was music, dancing, and pleasantry. He excelled in the last, and it attached everyone to him." It was 1759, and they both got out their violins and were soon playing the jigs and reels and other country dances that were part of the Virginia holiday celebration.

Patrick Henry wasn't happy as a shopkeeper weighing out flour, coffee, and sugar, so in his spare time he studied the law. In 1775—almost 16 years after that Christmas meeting—he said, "Give me liberty, or give me death," and helped inspire the American revolution.

But he never stopped fiddling. He was a country musician and always popular. Jefferson's taste was different. He loved serious music and amassed one of the largest collections of sheet (printed) music in America. He played the violin almost every day until the day he died.

Thomas Jefferson (above, in 1776, aged 33), designed and built his own folding music stand (right) for a string quartet.

13 A Massacre in Boston

John Adams hated to study when he was a boy. "I spent my time...in driving hoops, playing marbles...wrestling, swimming, skating."

Samuel Adams had a young cousin named John. "I have heard of one Mr. Adams," said King George to the Massachusetts governor, "but who is the other?" The other—honest, serious John Adams—would become even more famous than Sam.

Sam was an agitator and an organizer who helped start a revolution. John was a farmer and a lawyer, a solid citizen, who helped lead that revolution. Someone who knew John Adams said that he possessed more learning than anyone in the colonies. That may have been an exaggeration, but John had done a lot of reading and studying. And he knew how to use his mind.

Here is a story about both Adams cousins: the story of the Boston Massacre. A massacre, as you may know, is a gruesome killing. That's what happened in Boston in 1770.

The story begins in 1765, when the English Parliament passed a law that said American citizens had to provide quarters for British soldiers. The quarters they were talking about are not the kind you get when you add two dimes and a nickel. Quarters can also be houses where soldiers live. The law was called the Quartering Act. English soldiers, who were called redcoats because of the color of their uniforms, were to be quartered in American towns and cities.

Well, the Americans didn't want British redcoats quartered in their towns, or cities, or even in their country. So when the soldiers arrived, in 1768, the colonists weren't very kind to them. Sometimes they made fun of them, sometimes they threw snowballs or rocks, sometimes they called them lobsterbacks, or worse names.

PATRIOTIC AMERICAN FARMER.
D-K-NS—N, Esq; Barrister at La

Delaware's aristocratic John Dickinson wrote a series of articles that were made into a book, here titled *The Patriotic American Farmer*. He attacked British policy, but for different reasons from those that Tom Paine used. Dickinson said Parliament was acting radically and that it was the colonies that were trying to preserve ancient British liberties.

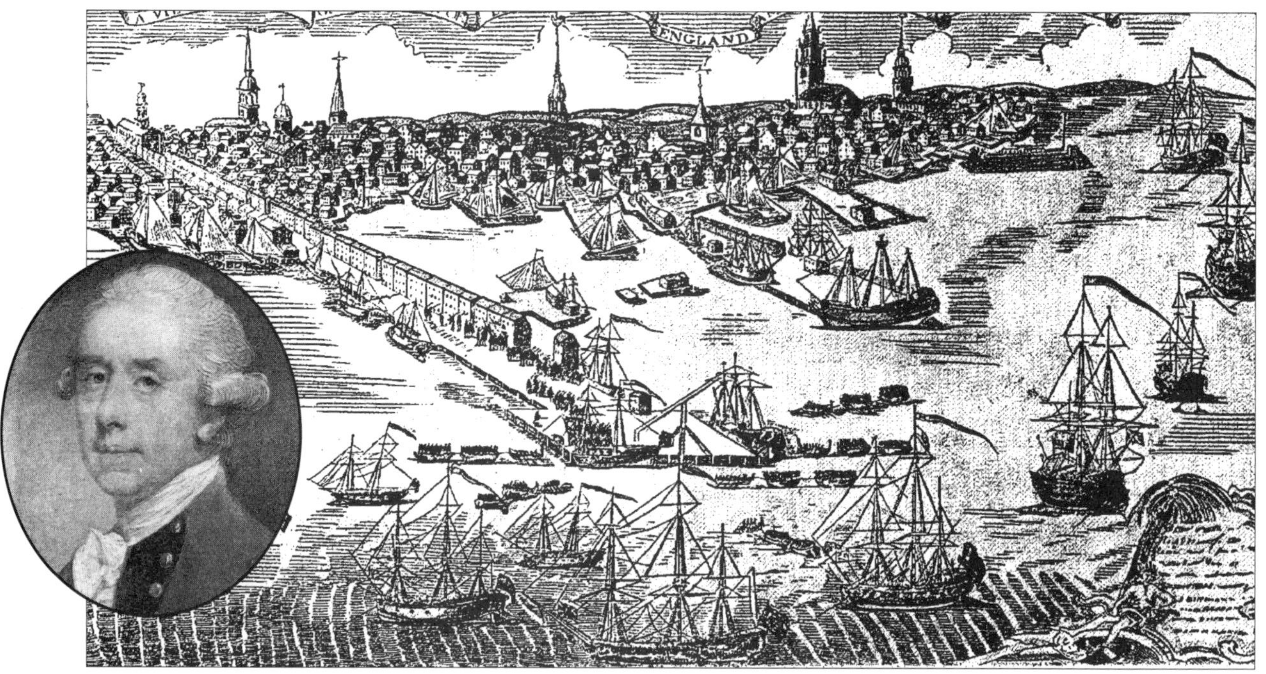

In 1768 the British fleet sailed into Boston harbor (above) and unloaded regiment after regiment of redcoats. General Thomas Gage (inset) had the thankless job of commanding an army that didn't want to be there among a people who didn't want it to be there either.

The people in Boston were especially annoyed, and, at first, wouldn't even provide quarters. So the soldiers set up tents on the Boston Common (a big grassy area in the center of town) and played their drums and bugles—loudly—at the most inappropriate times. Most of the English soldiers didn't want to be in America anyway. They were poorly paid, and many were homesick. Some ran away from the British army. (Soldiers who run away are called deserters. British deserters who were caught were shot.) A few redcoats—especially the officers—were treated well. Some married American women.

But for most of the British soldiers, the winters in Boston seemed longer and colder and more miserable than any they had ever known. On a freezing March day in 1770, one of the king's soldiers was looking for work to earn some extra money. Someone started making fun of him and told him to get a job cleaning toilets. (Only they didn't have the kind of toilets we have today. They had outdoor "privies," which were dirt-floored holes, and they smelled.) One thing led to another, and there was a fight.

That started things. Soon a noisy, jeering group of mischief-makers gathered in front of the Boston Custom House. They began pushing and shoving and throwing stones and pieces of ice at the British sentry. He got knocked down and he called for help. Captain Thomas Preston came to the rescue with eight British soldiers.

A particular Account of th most barbarous and

HORRID MASSACRE!

There is some confusion about what happened next. The mob is said to have taunted the redcoats, yelling "Fire! Fire!" Captain Preston is said to have yelled, "Hold your fire!" Then a British soldier was hit with a big stick. He claimed he heard the word "fire," so he fired his gun into the crowd. The street gang moved forward; the redcoats panicked and fired at unarmed people. Five Americans died; seven were wounded.

None of them was a hero. The victims were troublemakers who got worse than they deserved. The soldiers were professionals (the British army was supposed to be the best in the world), who shouldn't have panicked. The whole thing shouldn't have happened. Sam Adams made the most of it. He had Paul Revere engrave a picture of the shooting in front of the Boston Custom House. Adams was already calling it the Boston Massacre. Revere was a silversmith who made fine teapots and pitchers. He was also a dedicated Patriot, a dentist, a printer, a good horseback rider, and a friend of Samuel Adams.

Henry Pelham's drawing of the Boston Massacre (you can see Paul Revere's copy on the cover of this book). The coffins on the news sheet (top) carry the initials of the five Americans who died. Second from the right is C.A., for Crispus Attucks, a black laborer.

The picture that Paul Revere chose to etch into a piece of copper—so it could be printed over and over again—showed British soldiers firing at peaceful Boston citizens. That wasn't the way it had actually happened—Adams and Revere knew that—but the drawing made good propaganda. It made people furious at the British. That drawing was soon seen all over the colonies. It helped start a war.

There is one hero in the story of the Boston Massacre: John Adams. John didn't want British soldiers in Boston; he wanted freedom for his country. But he was fair and he always did what he thought was right. And even though everyone in America wanted to blame the British soldiers, John Adams believed they should have a fair trial. He knew they needed a good lawyer, and he was one of the best lawyers in the colonies. So he took

John Adams said that the representatives at the First Continental Congress (above) possessed "fortunes, ability, learning, eloquence, acuteness, equal to any I ever met with in my life."

A ***congress*** is a group of delegates who get together for discussion and action.

the case of the redcoats. Adams argued that the soldiers had defended themselves against an angry mob. A Boston jury found six of the soldiers not guilty. Two soldiers were found guilty of manslaughter—not murder; they were branded on their thumbs.

Long after the American Revolution, someone asked John Adams what the war had been about. There were two revolutions, he explained. One was the war itself. But the important revolution, he said, had occurred even before the war began. It had to do with ideas and attitudes. "The revolution was in the minds and hearts of the people," said John Adams. What do you think he meant by that?

John Adams was fighting for more than just separation from England. He wanted a chance to form a totally new kind of government: a government based on fair play and self-government.

John Jay (top) was a lawyer descended from two of New York's richest and most powerful families. Peyton Randolph (bottom) became the first president of the Congress.

Are people able to govern themselves? That question wasn't even being asked in most of the world. Always there were kings, or priests, or a ruling class. A country where people made their own laws? That sounded strange. Could the mass of people be trusted to choose their own leaders? It was a radical idea.

Samuel and John Adams knew that people in the colonies had much experience in self-government. They believed Americans could run their own nation and elect their own leaders. The Adams cousins would convince others; they would help form an American republic.

There was much to do before it could all work out. Plans had to be made. A congress was needed. Samuel Adams's Committees of Correspondence were made up of leaders from all of the colonies. Those committees then became a congress: the First Continental Congress.

It was 1774 when the congress met in Philadelphia, midway between New England and the Southern colonies. Philadelphia was America's leading city, so it made sense to meet there. Representatives came from every colony except Georgia. Samuel Adams and John Adams were both delegates. Sam wore a new wine-red suit with gold buttons, a gift from a Boston craftsman who didn't want his representative to look shabby. Alexander McDougall and John Jay (who would later be a new nation's first chief justice) came from New York determined to see that the colonies put pressure on England by not importing her goods. John Dickinson, who lived in Philadelphia, argued that a way must be found to get along with England. South Carolina's Christopher Gadsden and Virginia's Patrick Henry didn't agree with Dickinson. They were considered radicals. "Arms are a resource to which we shall be forced," said the fiery Patrick Henry. (When he said arms, he meant guns.) The Congress soon advised the colonists to form and arm militia (mill-ISH-uh) units and to stop buying goods from England.

Virginia's Peyton Randolph, a moderate, was elected president of the congress. South Carolina's John Rutledge (whom you first met when he

Remember, ***Patriots*** were Americans who wanted to be free of British rule. Sometimes Patriots were also called ***Whigs***—the Whigs were an English political party who mostly believed that the colonials should be allowed to govern themselves. Americans who supported the king were called ***Loyalists***—because they remained loyal to the existing government—and sometimes ***Tories***. The Tory political party believed the king should keep firm control of the colonies.

In 1774 King George wrote: "The New England governments are in a state of rebellion, blows must decide whether they are to be subject to this country or independent."

Historian Edmund S. Morgan says, "John Adams not only learned to work hard, but he always worked for something more than money.... He took the interesting cases rather than the lucrative ones and spent his time studying, studying, studying."

visited New York for the Stamp Act Congress) was another moderate. "There is in the Congress a collection of the greatest men upon this continent," John Adams noted in his diary.

The delegates at the Congress passed 10 resolutions listing the rights of the colonists, including the right to "life, liberty and property." But perhaps the most important thing that happened was that the colonial leaders got together and talked about their common problems. Then they wrote a polite, respectful petition and sent it to King George, urging him to consider their complaints. But George wouldn't even think about that.

The delegates made plans to meet again, if the situation in the country didn't improve.

Things got worse.

Knowing Where to Fight

Look at the map of Boston. Notice: Boston is a peninsula, a piece of land with a narrow neck connecting it to the mainland. (In the 18th century the neck was only a few hundred feet wide; today it is much wider.) Do you see how easy it would be to trap people in the city by stationing soldiers at the neck and putting boats in the harbor? The British weren't dumb. They figured that out. But, as you'll soon see, the Americans outsmarted them.

Now, check Charlestown, across the river. Paul Revere will begin a famous horseback ride in Charlestown. Notice Breed's Hill and Bunker Hill. A big battle will be fought on those two hills. The British will say they won the battle of the Hills—and they will capture them—but they will lose more than double the number of men the Americans lose. They can't afford many victories like that.

Do you see Dorchester Heights? Henry Knox will put cannons on that high spot above the harbor. You'll soon hear more about Knox and the trouble brewing in Boston harbor. Keep reading. The conflict is heating up.

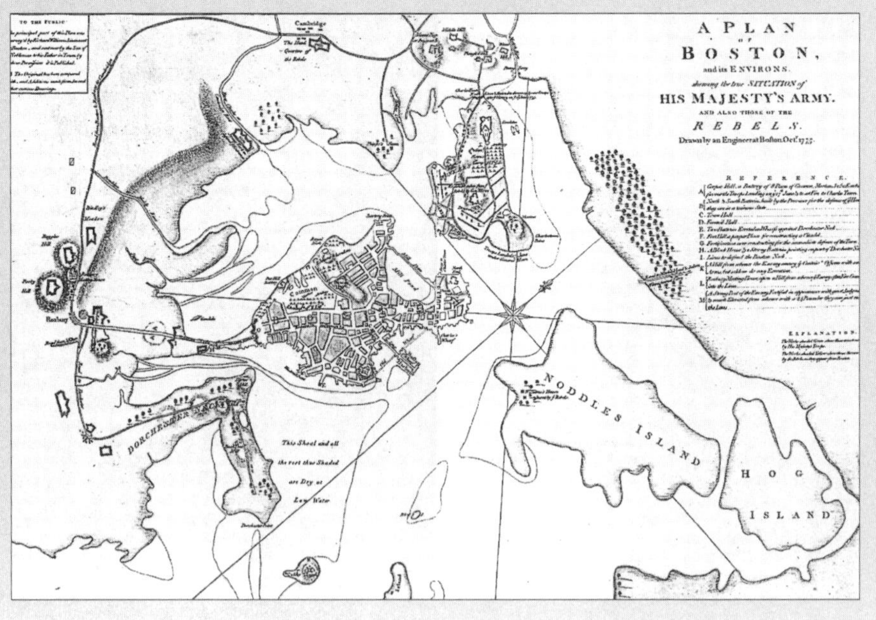

14 One If by Land, Two If by Sea

An American rifleman, better equipped than most of his fellow revolutionaries.

Three men rode horseback on an April night in 1775: Paul Revere, William Dawes, and Dr. Samuel Prescott. Each carried the same message: "The redcoats are coming." You may have heard of Paul Revere, because a poet, Henry Wadsworth Longfellow, wrote a famous poem about his ride.

Listen, my children, and you shall hear
Of the midnight ride of Paul Revere,
On the eighteenth of April, in Seventy-five;
Hardly a man is now alive
Who remembers that famous day and year.

Did you hear that Longfellow makes his words gallop, like horses' hoofs? If you read the story of Paul Revere in prose, you can compare it to the poet's version. Here it is:

The Patriots were worried. It looked as if war with Britain couldn't be avoided. The Patriots were the colonists who wanted independence. They wanted to be free of British rule. The other colonists—the ones who wanted to stay British subjects—were called Loyalists. Some Patriots, like Samuel Adams, expected war. But most Patriots still hoped to find peaceful ways to settle their differences with England.

It was scary to think of war. England was a great power; the colonies were scattered and had little military experience.

Still, it made sense to be prepared for the worst, so New Englanders began to stockpile cannonballs and gunpowder. They piled them up in Concord, a small town about 20 miles northwest of Boston.

When the British officers heard about those munitions, they decided to get them. Paul Revere and his Boston friends learned of the

TO all Gentlemen VOLUNTEERS, who prefer LIBERTY to SLAVERY, and are hearty Friends to the GRAND *AMERICAN* CAUSE; who are free and willing to serve this STATE, in the Character of a Gentleman MATROSS, and learn the noble Art of Gunnery, in the Massachusetts State Train of Artillery, commanded by Col. THOMAS CRAFTS, now stationed in the Town and Harbour of *BOSTON*, and not to be removed but by Order of the honorable House of Representatives, or Council of said State; let them appear at the Drum-Head, or at the [illegible] where [illegible] shall enter into present Pay [illegible] at [illegible] *Shillings per Month*. For their Encouragement they shall receive *Twenty Dollars* Bounty on passing Muster, one Suit of Regimental Cloathes yearly, a Blanket, &c. with Arms and Accoutrements suitable for a Gentleman Matross. For their further Encouragement, the Colonel would inform all Gentlemen Volunteers, that there are *twenty-two Non-commission Officers in each Company, who receive from three Pounds four and six Pence, to three Pounds twelve per Month*; and as none will be accepted in said Regiment, but Men of good Characters, such only will be promoted, whose steady Conduct and good Behaviour merits it.

You are desired to take Notice of the difference of Pay and Station.

Advertisements like this one, for volunteers to fight for the colonies, were soon plastered around Boston.

British plans. They found out that the redcoats were going to march on Concord the next morning!

Besides the gunpowder, Revere was worried about his friends Sam Adams and John Hancock. They were hiding in Lexington, right next door to Concord. The British were searching for those two troublemakers—they wanted to put them in jail.

Someone had to get a warning to those towns—and fast. It would

Redcoats march through Concord looking for weapons and ammunition while their officers spy out the land. Too late.

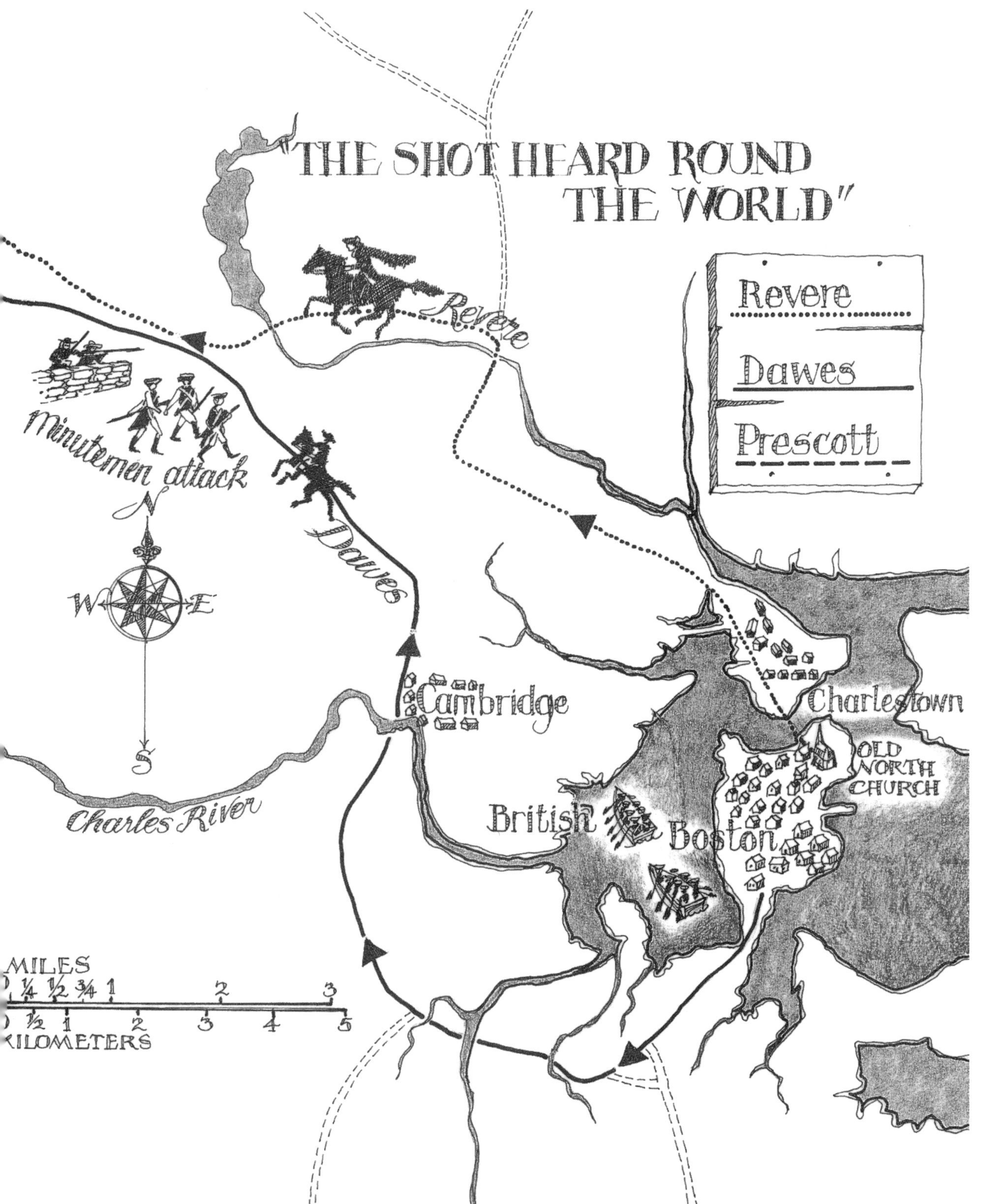
"THE SHOT HEARD ROUND THE WORLD"
Revere
Dawes
Prescott
Revere
Minutemen attack
N
W
E
S
Dawes
Cambridge
Charlestown
OLD NORTH CHURCH
British
Boston
Charles River
MILES
0 1/4 1/2 3/4 1 2 3
0 1/2 1 2 3 4 5
KILOMETERS

Who Started It?

The story of Lexington and Concord as seen in Massachusetts and in England:

The troops came in sight just before sunrise...the Commanding Officer accosted the militia in words to this effect: "Disperse, you rebels, damn you, throw down your arms and disperse;" upon which the [American] troops huzzaed, and immediately one or two [British] officers discharged their pistols, which were instantaneously followed by the firing of four or five of the soldiers, and then there seemed to be a general discharge from the whole body. Eight of our men were killed and nine wounded.

—FROM THE *SALEM GAZETTE*, SALEM, MASSACHUSETTS, APRIL 25, 1775

Six companies of light infantry...at Lexington found a body of the country people under arms, on a green close to the road. And upon the King's troops marching up to them, in order to inquire the reason of their being so assembled, they went off in great confusion. And several guns were fired upon the King's troops from behind a stone wall, and also from the meeting house and other houses....In consequence of this attack by the rebels, the troops returned the fire and killed several of them.

—FROM THE *LONDON GAZETTE*, LONDON, ENGLAND, JUNE 10, 1775

A ***green*** is a grassy lawn or common. Many New England villages have a green for public gatherings.
To ***accost*** someone means to approach and speak to or touch him or her—but not gently.

Disperse means to break up and scatter.
Huzza is an old-fashioned word for "yell." It's something like "hurrah." The rebels were yelling at the British soldiers.

help to know which way the redcoats would march. Would they go by the long land route over the Boston neck? Or would they take the shorter route—by boat across the water to Charlestown and then on foot from there?

Billy Dawes didn't wait to find out. He pretended to be a drunk farmer and staggered past the British sentry who stood guard at the neck. As soon as he was out of sight of the guard, Dawes jumped on a horse and went at a gallop. He knew the redcoats would start out soon, and he shouted that message at each Patriot house he passed.

That same dark night Paul Revere sent someone to spy on the British. "Find out which way the redcoats will march," the spy was told. "Then climb into the high bell tower of the North Church and send a signal. Light one lantern if they go by land. Hang two lanterns if they go by sea."

Revere got in a boat and quietly rowed out into the Charles River. A horse was ready for him on the Charlestown shore. He waited—silently. (Revere was a known Patriot and would have been arrested if the British had found him outdoors at night.)

Grenadiers, dragoons, regulars, redcoats—they're all British soldiers.

And lo! as he looks, on the belfry's height
A glimmer, and then a gleam of light!
He springs to the saddle, the bridle he turns,
But lingers and gazes, till full on his sight
A second lamp in the belfry burns!

Now he knew! The redcoats would take the water route across the Charles River, just as Paul Revere was doing. What happened next? Well, both Billy Dawes and Paul Revere rode hard, through the night, warning everyone in the countryside that the British were coming. They met at Lexington in time to tell Sam Adams and John Hancock to escape. But before they could go on to Concord, they were stopped by a British patrol. The redcoats took their horses. Luckily, by this time, a third man, Dr. Samuel Prescott, was riding with Dawes and Revere. (Prescott had been visiting the girl he intended to marry, who lived in Lexington.) The doctor managed to escape from the British, ride home to Concord, and warn everyone there.

The American farmers were ready, and they grabbed their guns. They were called minutemen because they could fight on a minute's notice. (Some had been trained fighting in the French and Indian War.) Captain John Parker was the leader of the minutemen, and what he said on that day is now carved in stone near the spot where he must have stood. "Stand your ground. Don't fire unless fired upon. But if they mean to have a war let it begin here!"

The stanza with Ralph Waldo Emerson's famous line goes like this:

By the rude bridge that arched the flood,
Their flag to April's breeze unfurled,
Here once the embattled farmers stood,
And fired the shot heard round the world.

And it did begin right there, at Lexington. Each side said the other fired the first shot. No one knows who really did, but a poet named Ralph Waldo Emerson called it "the shot heard round the world." (Can you see why?)

When the smoke cleared, eight American farmers lay dead. It was April 19, 1775. The American Revolution had begun.

But it was gunpowder that the redcoats had set out to get, so they marched on—to Concord—but they couldn't find the powder. That made them so angry they started a fire. "Will you let them burn the town down?" shouted one colonist. "No, I haven't a man who is afraid to go,"

Senior Citizen's Arrest

One of the best stories of the Revolutionary War involves an old woman. She was called Mother Batherick, and she lived near Concord. On the day of the famous battle, all the young men of her town became minutemen and rushed off to fight, leaving behind a group of old men. They were supposed to guard the town. The old soldiers chose as their leader a veteran of the French and Indian War, a black man whose name was David Lamson. Lamson and his men were all behind a stone wall when some British supply wagons came by. Lamson told the redcoats to halt. They didn't, and the old warriors fired. Two British soldiers and four horses went down. The other redcoats ran.

Mother Batherick was digging weeds at a nearby pond. Six breathless British soldiers rushed up and surrendered to her. She turned them over to Lamson and his old troopers. After that, Americans liked to ask this question: "If one old lady can capture six grenadiers, how many soldiers will King George need to conquer America?"

Paul Revere in his regular job as a silversmith, showing off one of his teapots.

said the minutemen's Captain Isaac Davis. The British stood at the North Bridge in Concord. They fired at the colonists. The minutemen fired back. Now the British were scared, and they tried to retreat. The Americans followed and whipped the redcoats. More than two Englishmen fell for every American casualty.

Do you know the song "Yankee Doodle"? Well, the British made it up to insult the Americans. They said a Yankee Doodle was a backwoods hick who didn't know how to fight. When the British marched to Concord and Lexington, they wore their fancy red uniforms, and their drummers and pipers played "Yankee Doodle."

After the battle, it was the Americans who sang that song. They said, "We'll be Yankee Doodles and proud of it!"

But that isn't the whole story. There is always more to war than winning or losing. These are words written in 1775:

> *Isaac Davis...was my husband. He was then thirty years of age. We had four children; the youngest about fifteen months old....The alarm was given early in the morning, and my husband lost no time in making ready to go to Concord with his company...[he] said but little that morning. He seemed serious and thoughtful; but never seemed to hesitate....He only said, "Take good care of the children." In the afternoon he was brought home a corpse.*

Yankee Doodle

Yankee Doodle went to town,
A-ridin' on a pony.
Stuck a feather in his cap
And called it Macaroni.

Chorus:
Yankee Doodle, keep it up,
Yankee Doodle Dandy,
Mind the music and the step
And with the girls be handy.

Father and I went down to camp,
Along with Captain Gooding,
And there we saw the men and boys
As thick as hasty pudding.

(Chorus)

And there we saw a thousand men,
As rich as Squire David;
And what they wasted every day,
I wish it could be savéd.

(Chorus)

And there was Captain Washington
Upon a slapping stallion,
A-giving orders to his men;
I guess there was a million.

(Chorus)

And there I saw a little keg,
Its head was made of leather;
They knocked upon it with two sticks
To call the men together.

(Chorus)

And there I saw a swamping gun,
As big as a log of maple,
Upon a mighty little cart,
A load for father's cattle.

(Chorus)

And every time they fired it off
It took a horn of powder,
It made a noise like father's gun,
Only a nation louder.

(Chorus)

I can't tell you half I saw,
They kept up such a smother,
So I took my hat off, made a bow
And scampered home to mother.

(to tune of chorus)
Yankee Doodle is the tune
Americans delight in.
'Twill do to whistle, sing or play
And just the thing for fightin'.

After Lexington and Concord, a known British sympathizer—a Loyalist—could be strung up and ridiculed, like this man, or sometimes find a worse fate.

Henry Wadsworth Longfellow finished the story:

You know the rest. In the books you have read
How the British Regulars fired and fled,—
How the farmers gave them ball for ball,
From behind each fence and farmyard wall,
Chasing the redcoats down the lane,
Then crossing the fields to emerge again
Under the trees at the turn of the road,
And only pausing to fire and load.

So through the night rode Paul Revere;
And so through the night went his cry of alarm
To every Middlesex village and farm,—
A cry of defiance and not of fear,
A voice in the darkness, a knock at the door,
And a word that shall echo forevermore!
For, borne on the night-wind of the Past,
Through all our history, to the last,
In the hour of darkness and peril and need,
The people will waken and listen to hear
The hurrying hoof-beats of that steed,
And the midnight message of Paul Revere.

The people of New England did not wish for war. This was not a warrior culture...and showed none of the martial spirit that has appeared in so many other times and places. There were no cheers or celebrations when the militia departed. ...The people of New England knew better than that. In 140 years they had gone to war at least once in every generation, and some of those conflicts had been cruel and bloody. Many of the men who mustered that morning were themselves veterans of savage fights against the French and Indians. They and their families knew what war could do.

—DAVID HACKETT FISCHER, *PAUL REVERE'S RIDE*

15 An American Original

The Patriots took Ticonderoga on May 10, 1775, the very day the Second Continental Congress convened.

This statue of Ethan Allen has disappeared. It was the only portrait said to be a good likeness by men who knew him.

Ethan Allen was born in Connecticut in 1738—that much we know as fact. The stories say he was asked to leave his hometown. It was his tongue that got Ethan Allen in trouble. His language was rough and rowdy, more than fine-mannered Connecticut citizens could handle.

They couldn't put a muzzle on Ethan Allen. In fact, they couldn't do much of anything with him. He was a sinewy giant of a man, famous for his strength. Like George Washington, he could outwrestle and out-throw and outlift any challengers. But he may not have felt like fighting all of Connecticut.

So off he went to the New Hampshire Grants to farm and speak his mind. The New Hampshire Grants was the green mountain land between New Hampshire and New York that is now Vermont. New Hampshire sold the land to farmers; then England said the land belonged to New York. New York told the farmers to move out or buy the land again.

Now no one could tell Ethan Allen what to do. He and his friends banded together, calling themselves the Green Mountain Boys. They boasted that they could shoot a nut out of the jaws of a squirrel, and perhaps they could. But it wasn't squirrels they went after. It was the "Yorkers" who were moving into their territory.

Then the revolution came, and the Green Mountain Boys had something else to fight about. It was at the Catamount Tavern, in the little town of Bennington, in 1775, that Ethan Allen had a famous meeting

A ***catamount*** is a big cat with many names. The word is also used to describe the ***puma***, the ***panther***, the ***cougar***, and the ***mountain lion***. Catamounts live on birds and small animals (which makes them ***predators***). These felines used to roam most of the northern United States, especially the Rocky Mountain area.

with Patriots from Connecticut. Allen liked the tavern. It had a big stuffed catamount that snarled in the direction of New York.

The Patriots had come to ask the Green Mountain Boys for help. They brought some fighters and some cash, and they paid for the drinks. Allen and his men agreed to do what they asked. They agreed to capture an important British fort—Fort Ticonderoga—on Lake Champlain.

Along with bluster and strong muscles, Ethan Allen also had a good mind. He planned the attack carefully. First he sent a spy. The spy was a farmer who wandered into the British fort pretending he was looking for a barber. What he was really looking for was information. He found what he came for. The spy told Allen the number of men in the fort. Then he told him there was a weak spot in the walls where the fort could be entered.

Just as Allen and the Green Mountain Boys were getting ready to start out, Benedict Arnold appeared.

Arnold was a colonel in the Continental army, with a fancy uniform and a valet. He later became a traitor, going over to the British side. Despite that villainy, it is only fair to tell you he was a strong leader and a good fighter. Benedict Arnold said he was going to take command of the mission, under orders from Massachusetts.

Ethan Allen couldn't stand Arnold; he called him a "damned rascal." They might have fought each other, but time was running out and the attack was set for early morning. So they agreed to lead together, shoulder to shoulder.

Into the fort they went—Benedict Arnold like a proper soldier and Ethan Allen and his men howling war whoops like Mohawk Indians. A startled sentry ran for his life.

Then a British officer appeared with his pants in his hands—he hadn't had time to put them on. The astonished officer asked on whose authority they attacked. Ethan Allen roared an answer that was to become famous: "In the name of the Great

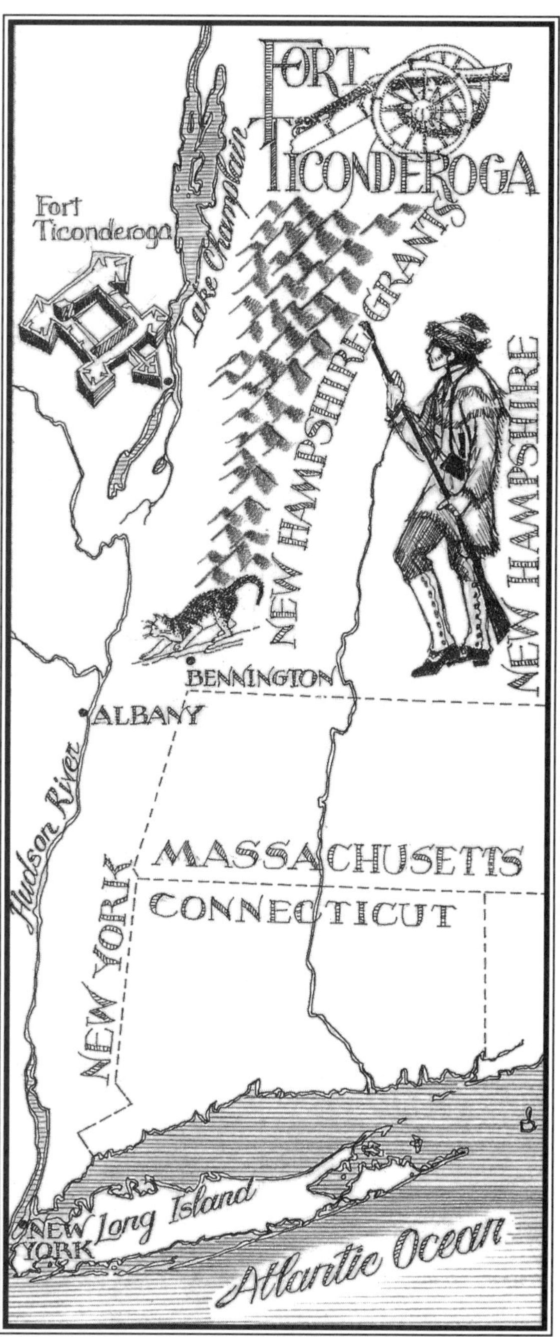

Fort Ticonderoga guarded Lake Champlain. Now the Americans held the route into Canada.

Benedict Arnold was a good soldier and very brave in battle. He once said about himself that "his courage was acquired, and he was a coward till he was 15 years old." It was a terrible shock to the Patriots when he went over to the British side in 1780.

Jehovah and the Continental Congress!"

It wasn't much longer before the fort, its cannons, and its rum all belonged to the Patriots.

No one was killed. That was typical of Ethan Allen. He was a curser, not a killer. He'd capture Yorkers or redcoats and beat them with a beech stick, but he wouldn't shoot them.

Four months later Ethan Allen led a force to Canada; he planned to capture Montreal—or so he said. Instead, he was captured, and he spent two years in prison. When he got out of jail, he met General Washington. Washington said of him, "There is an original something in him that commands attention."

Original he was. He had his own ideas on religion and God, and he wrote a book explaining his beliefs. He helped establish Vermont as an independent republic. If New York and New Hampshire and even Massachusetts couldn't agree on Vermont's land, Vermont would just go its own way. And it did, until 1791, when it became the 14th state.

Ethan Allen, waving his sword on the right, catches the British commander at Fort Ticonderoga (second from right)—with his pants down.

Vermont was an independent nation from 1777 to 1791. It coined its own money, ran its own postal service, had its own constitution, and was the first state to outlaw slavery.

How the New World Changed the Old, and Vice Versa

"See the great Newton, he who first survey'd,
The Plan by which the Universe was made."

Lights were being lit in the 18th century—so many lights that it would come to be called a time of "Enlightenment." The lights were going on in the minds of the thinking people. Some of the electricity for those lights had come from a scientist named Isaac Newton. Newton had shown that the universe was not as full of mystery as people had supposed. It could be understood with study and observation and by people using their brains. That was an astonishing thought in a world that had often been guided by superstition and fear. Suddenly there seemed to be all kinds of brilliant thinkers who were using their minds and encouraging others to do the same thing.

An Englishman named John Locke and a Frenchman named Jean-Jacques Rousseau (jahn-jahk-roo-SO) were two of the most important Enlightenment thinkers. They thought about politics and the way governments were run. They got some of their ideas by considering the American Indian and the "New World." Jean-Jacques Rousseau called Indians "noble savages." John Locke said, "In the beginning, all the world was America." (What do you think he meant by that?)

John Locke wrote about natural rights. He said that governments should be run for the people, not for their rulers. Locke made people think about democracy.

The colonists in America read what Locke wrote. They read about the ancient democracies in Greece and Rome. They knew that most American Indians seemed to live a free, democratic life in self-governing tribes.

Philosopher John Locke wrote, "Wherever Law ends, Tyranny begins."

The colonists knew something else: they knew they could govern themselves. They didn't need kings or nobles to make decisions for them. Americans had been running their own assemblies for years. There was the General Court in Massachusetts, the House of Burgesses in Virginia, and law-making bodies in each colony. Nowhere in Europe did people have that kind of experience in self-government.

An article published in England around 1776 said, "The darling passion of the American is liberty and that in its fullest extent; nor is it the original natives only to whom this passion is confined; our colonists sent thither seem to have imbibed the same principles."

"Never exceed your rights," wrote Rousseau, "and they will soon become unlimited." What did he mean?

Americans were sending raw materials to England—like lumber and tobacco—and getting them sent back as finished goods—furniture and cigars. Well, another raw material got sent back and forth across the sea: the idea of freedom and democratic government.

16 On the Way to the Second Continental Congress

Do you think we should make heroes of our politicians? What kinds of people might want to serve in Congress if we treated politicians like superstars? Do you think politicians get enough attention? Or too much?

A Virginian described Washington as "sensible, but speaks little." Washington spoke up when something mattered. In 1785 he wrote about slavery: "There is not a man alive who wishes more sincerely than I do, to see a plan adopted for the abolition of it." He believed slavery should be ended by "legislative authority" (laws), because slave owners would not willingly give up wealth and property. Washington remained a slave owner himself. He freed his slaves in his will.

Washington, said his friends, was serious but never stern, and always cheerful with his soldiers.

Pretend it is 1775. You are a British subject living in the American colonies in Philadelphia. At least that is the way you have been taught to describe yourself. But now you are confused. You have overheard violent arguments. Some people are calling the Bostonians "heroes"; others call them "rabble." Politics is making people angry. Your parents are no longer talking to some of their old friends.

Your parents are Patriots; some of your neighbors are Loyalists. If there is war, the Loyalists hope Britain will win. They don't see any need for independence. England is the greatest nation on earth, they say. They remember the good old days before the French and Indian War. England didn't bother the colonists with many taxes then. They expect those times to return again. Benjamin Franklin's son William is a Loyalist. He is sincere in his beliefs, but he will break his father's heart.

Being a Patriot may mean going to war. That worries you—and it should. What side will you be on? In May, when the Virginia delegation arrives in Philadelphia, you make a decision. You will stick with the American Patriots' cause.

Back in the 18th century there were no TV stars and no big sports figures, which may explain why, in 1775, everyone in Philadelphia seemed to want a glimpse of Virginia's political leaders when their carriages rolled into town. The Virginians had been in Philadelphia the year before, when the First Continental Congress met. Now they were back for the Second Congress: heroic-looking men who rode their horses proudly, who danced with energy and grace, and who thought and spoke as well as any Americans anywhere. Even John Adams of Massachusetts said that they represented "fortunes, ability, learning, eloquence, acuteness, equal to any I ever met with in all my life."

And So to Bed

Dr. Benjamin Franklin

In early America, inns were often crowded, and travelers expected to share beds. It happened to Ben Franklin and John Adams one night in 1776, when "but one bed could be procured for Dr. Franklin and me in a chamber little larger than the bed." Adams, with his fussy ways, wanted to close the window. (Most physicians then thought the night air foul and dangerous.) This is what Adams wrote in his diary about Franklin's views:

Oh! says Franklin, don't shut the window. We shall be suffocated. I answered I was afraid of the evening air. Dr. Franklin replied...come! open the window...and I will convince you....Opening the window and leaping into bed, I said I had read his letters to Dr. Cooper...but the theory was so little consistent with my experience that I thought it a paradox. However, I had so much curiosity to hear his reasons that I would run the risk of a cold. The Doctor then began a harangue upon air and cold and respiration and perspiration, with which I was so much amused that I soon fell asleep, and left him and his philosophy together.

Take George Washington, for instance. He was more than six feet tall, big-boned, muscular, lean, and very strong. Once he came upon some young men who were throwing weights as far as they could. They had their shirts off and were sweating from the effort. George Washington asked if he could try. He took a weight—didn't even take off his jacket—and out-threw them all. Does that sound as if he was a show-off? He wasn't. Everyone agreed about that. He was modest, and only spoke when he had something to say.

His adventures during the French and Indian War had made him famous, even in England. In America both men and women admired him. One friend called him "the best horseman of his age and the most graceful figure that could be seen on horseback." He had gray-blue eyes, auburn hair, and hands and feet so large that several people of his time remarked about them. He loved to dance and he dressed with care. He wore his military uniform to Philadelphia—bright blue with brass buttons—and they called him Colonel Washington.

When John Adams's wife, Abigail, met George Washington she found a poem to describe him:

Mark his majestic fabric; he's a temple
Sacred by birth, and built by hands divine.

George Washington was in his coach, riding to Philadelphia with

Stephen Hopkins of Rhode Island had "palsy," which was a vague term covering many illnesses in the 18th century. Whatever his disability, it didn't keep him out of a long career in public service. It didn't limit his enthusiasm for independence, either. As he put his pen to the paper to sign the Declaration, he said proudly: "My hand trembles, but my heart does not."

Step. Hopkins

another Virginian: Richard Henry Lee. The fingers on one of Lee's hands had been shot off in a hunting accident; he kept a silk handkerchief wrapped around that hand and pointed with it when he spoke. That should give you an idea of the man's style. He was good-looking and wore elegant clothes, and he talked smoothly.

Richard Henry Lee (above) and Patrick Henry were Virginia's best speakers.

Lee was full of surprises. He was a slave owner, but he hated slavery and spoke out against it. Though he was dashing and aristocratic, he got along well with rumpled Samuel Adams. It was Richard Henry Lee (along with Patrick Henry and Thomas Jefferson) who organized the first Committee of Correspondence in Virginia.

Lee came from a talented family. His brothers were all outspoken leaders. That means they said what they believed. So did his sister Hannah. She was furious when she was turned away from the voting polls because she was female. It was taxation without representation, said Hannah Lee.

As Lee and Washington rode toward Philadelphia, they were joined

Meet Some of the Delegates

Delegate PHILIP LIVINGSTON lived like a prince in New York. His family had been prominent in the colonies for five generations, but Philip Livingston made his own fortune as a trader and privateer during the French and Indian War. In spite of his wealth, he identified with ordinary people and opposed the colony's royal governor and the Stamp Tax. Livingston believed in political and religious freedom.

JOSEPH HEWES, who came from North Carolina, was opposed to separation from Great Britain—even when people in North Carolina told him to vote for it. Then, during a debate at the convention, something happened. "He started suddenly upright," reported John Adams, "and lifting up both his hands to Heaven, as if he had been in a trance, cried out, 'It is done! and I will abide by it.'" Hewes was now for independence!

STEPHEN HOPKINS, who was elected governor of Rhode Island 10 times, attended the Albany Congress in 1754 with Benjamin Franklin, Sir William Johnson, and Hendrick. Stephen Hopkins helped Ben Franklin write a plan for a union of the colonies. Most Americans weren't ready for that in 1754. Now it seemed that they were.

Georgia's BUTTON GWINNETT had an unforgettable name and just a year to live. Gwinnett—Georgia's governor—was killed in a duel. Afterwards, no one could remember what the duel was about—except honor, they said.

by other members of the Virginia delegation. Farmers along the way took off their hats and cheered. Then, six miles from Philadelphia, 500 soldiers on horseback appeared to escort them. By the time they entered the city, a military band was playing and infantrymen were marching—it was some parade.

The Virginians were the same seven men who had been at the first congress in 1774 (although some would leave almost immediately and others would take their place). Three were the best orators in the state, perhaps in the nation: Patrick Henry (who looked like a country boy, and seemed to want it that way), Richard Henry Lee (who asked this congress to declare for independence), and slim, graceful Edmund Pendleton (who debated with cool logic).

Virginia's Benjamin Harrison was the biggest man at the Convention. He was six feet four inches tall and was said to weigh 400 pounds. (Many of the delegates were big—it was normal to be heavy. Meals were large: soup, fish, meat, vegetables, potatoes, pie and cake, fruit and cheese—all at one sitting. John Adams, just five feet six inches tall, grew to weigh 275 pounds.) Harrison told a friend he would have come to this convention on foot, if he'd had to, rather than not come. He became governor of Virginia; his son and great-grandson became presidents of the United States.

Popular Peyton Randolph, another giant of a man, had been president of the First Continental Congress and was expected to preside again. But he did not stay long. Nor did Patrick Henry. They were needed in Williamsburg. Virginia's House of Burgesses had been called back into session. State business seemed more important to them than anything that might occur at this experimental gathering.

Peyton Randolph's cousin, who was just 33, came to take his place in Philadelphia. The cousin was a thoughtful, quiet man who was known to be a good writer. His name was Thomas Jefferson.

The Virginians were the crowd-pleasers, but the congress as a whole was so extraordinary it would still inspire awe 200 years later.

The Adams cousins—Sam and John—were back from Massachusetts, along with rich John Hancock, who became president of this Second Continental Congress. John Witherspoon, a Scotsman who had needed persuading to come to America to head Princeton College, was a delegate from New Jersey. So was Francis Hopkinson, an inventor and scientist who wrote poetry, composed music, and painted.

Of Caesar Rodney, the delegate from Delaware, John Adams wrote: "[He] is the oddest looking man in the world; his face is not bigger than a large apple, yet there is a sense of fire, spirit, wit, and humor in his countenance."

Virginia's Benjamin Harrison when he was young and slim.

Thomas Jefferson was President Peyton Randolph's cousin.

Francis Hopkinson, a delegate from New Jersey.

Men of the Middle Colonies

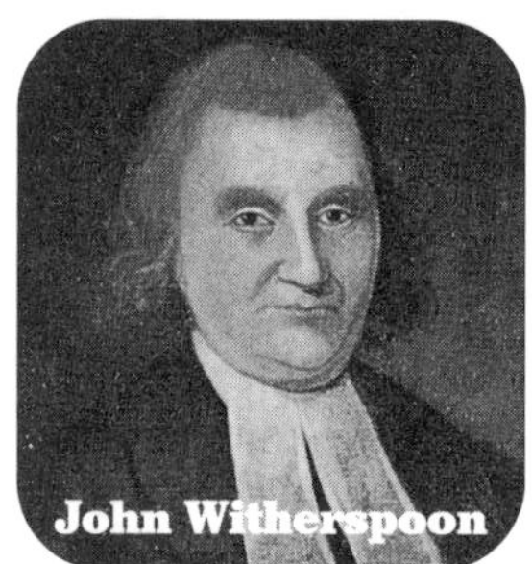

Benjamin Rush went on to serve as an army surgeon during the Revolutionary War. He set up the first free clinic in America, and became the country's most famous medical professor. When Rush was studying medicine in Edinburgh, Scotland, he helped persuade John Witherspoon, a famous Scots clergyman, to come to America to be president of Princeton, where Rush had gone to college. Dr. Witherspoon was the only minister to sign the Declaration of Independence. Charles Carroll of Maryland was the last of the Signers to die (in 1832) and the only Roman Catholic one.

Benjamin Rush was a doctor and a teacher. He'd learned medicine as an apprentice to a doctor and then had gone to Scotland to learn more. Rush had ideas that seemed strange to some people: He hated slavery, tobacco, and capital punishment. He thought girls and blacks should go to school and that they could learn as much as white boys. Rush was one of Pennsylvania's representatives, and remarkable. Pennsylvania's Ben Franklin was even more so.

No American was better known than Benjamin Franklin. He'd come to Philadelphia from Boston as a penniless boy and soon made his fortune as a printer and publisher. He made his fame as an inventor, scientist, philosopher, and political leader. Franklin had spent years in London as an agent for several of the colonies. No one tried harder than he to avoid a break with England. He proposed the idea of a British commonwealth of independent nations, each with its own parliament, but all with the same king. The leaders of Britain's Parliament rejected that idea and treated Franklin with contempt.

Franklin changed his thinking; he began to favor independence. He arrived home from England on May 5, 1775, just in time to attend the opening of this congress.

In March he was traveling again, this time on a wild goose chase to Canada. It was hoped that the Canadians would join the other colonies and fight Britain. Franklin and two delegates from Maryland—Charles Carroll (said to be the wealthiest man in America) and Samuel Chase (who was a leader of the Sons of Liberty in Annapolis)—headed north together. It was an exhausting trip, especially for 69-year-old Benjamin Franklin. (In Albany they noted that most people still spoke Dutch. In upper New York, they had to sleep in the snowy woods.) When they finally arrived at their destination, they couldn't persuade the Canadians to join the Revolution. (Religion had something to do with it. Catholic Canada feared an alliance with the mostly Protestant colonies.)

In June Franklin was back at the convention, where he was asked to serve (with John Adams and Thomas Jefferson) on a committee that was to write an important declaration. Some people say this was the most important political statement ever written. It was addressed to King George III. Hold on for a few chapters and I'll tell you all about it.

17 Naming a General

John Adams wrote, "I am determined this morning to make a direct motion that Congress should adopt the army before Boston."

At first the Continental Congress found itself in a strange situation. Americans were in fighting mood, but war had not been declared. Should they prepare for war? Should they work for peace? Could they do both?

People were calling for a Continental army. The minutemen who fought at Lexington and Concord were gathered near Boston. Others had come from the countryside with rifles and muskets. If someone didn't take charge they would all go back home.

The Continental Congress couldn't ignore the problem, especially after a letter arrived from the Boston Patriots pleading for the Congress to take over their forces.

John Adams spoke up. He called for a "Grand American Army" to be made up of volunteers from all of the colonies. The guns fired at Lexington and Concord might be heard next in Charleston, or Baltimore, or even in Philadelphia, Adams told the delegates. They must have shuddered, because they knew he spoke the truth.

In each of the colonies, citizen soldiers—militia—were ready to fight. Someone had to organize the militias and the minutemen into an army. A general was needed, said Adams.

John Hancock from Massachusetts believed he was the man for the job. He had done a bit of soldiering, and it was his money that was paying some of Congress's bills. So when John Adams stood up to nominate a general, almost everyone—especially John Hancock—thought it would be Hancock. But, as you know, John Adams always did what he

1775: Making a Revolution

April 19: the battle of Lexington and Concord.

May 10: Second Continental Congress convenes in Independence Hall, Philadelphia.

May 10: Ethan Allen and Benedict Arnold capture Fort Ticonderoga.

June 15: George Washington appointed head of the Continental army.

June 17: the Battle of Bunker and Breed's Hills.

July 3: General Washington takes command of 17,000 men at Cambridge, Massachusetts.

July 26: the Continental Congress establishes a post office department and appoints Franklin Postmaster General.

August 1: Tom Paine publishes an article in the *Pennsylvania Gazette* supporting women's rights.

August 23: George III declares the American colonies in rebellion.

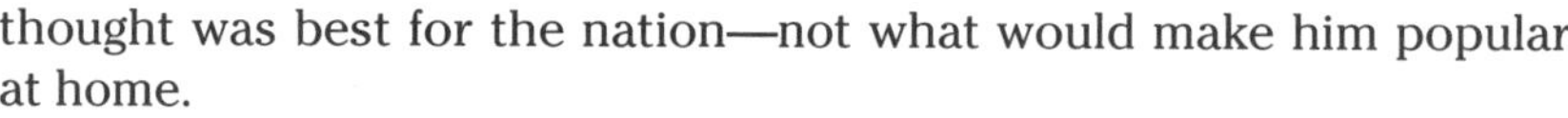

As soon as the troubles with England were settled, the citizens of Charles Town, South Carolina, officially changed their city's name to Charleston. The capital was moved from there to Columbia.

thought was best for the nation—not what would make him popular at home.

"There is but one man in my mind for this important command," said Adams, and Hancock looked pleased. "The gentleman I have in mind…is from Virginia." When Adams said that, John Hancock's face fell, and Washington, who realized he was the man from Virginia, rushed from the room.

John Adams continued, "[His] skill as an officer…great talents and universal character would command the respect of America and unite …the Colonies better than any other person alive." The congressmen agreed. George Washington was elected general unanimously.

He accepted—on one condition. He would take no salary. And that was part of Washington's greatness. He was willing to serve without pay for a cause he thought noble.

Washington knew that the general's job could lead to disaster. England was the greatest power in the world. Its army was well-trained and supplied with the latest guns and cannons. Its navy was the finest in the world.

The American army was made up of a raggedy bunch of men—farmers, shoemakers, carpenters, blacksmiths—who had few

John Hancock (left) was disappointed not to be put in charge of the army. But as president of the Second Continental Congress he signed the order naming Washington commander in chief (below).

In Congress

The delegates of the United Colonies of New hampshire, Massachusetts bay, Rhode-island, Connecticut, New-York, New-Jersey, Pennsylvania, New Castle Kent & Sussex on Delaware, Maryland, Virginia, North Carolina & South Carolina

To George Washington Esquire

We reposing special trust and confidence in your patriotism conduct and fidelity Do by these presents constitute and appoint you to be General and Commander in chief of the army of the United Colonies and of all the forces raised or to be raised by them and of all others who shall voluntarily offer their service and join the said army for the defence of American liberty and for repelling every hostile invasion thereof And you are hereby vested with full power and authority to act as you shall think for the good and welfare of the service And we do hereby strictly charge and require all officers and soldiers under your command to be obedient to your orders & diligent in the exercise of their several duties And we do also enjoin and require you to be careful in executing the great trust reposed in you by causing strict discipline and order to be observed in the army and that the soldiers are duly exercised and provided with all convenient necessaries And you are to regulate your conduct in every respect by the rules and discipline of war (as herewith given you) and punctually to observe any such orders and directions from time to time as you shall receive from this or a future Congress of the said United Colonies or a Committee of Congress for that purpose appointed

This Commission to continue in force until revoked by this or a future Congress

Dated Philadelphia June 19th 1775

By order of the Congress John Hancock President

Attest. Cha Thomson secy

Immediately after he was named General, Washington went to Cambridge, Massachusetts, to take charge of the Continental army (left).

The Congress did pay Washington's expenses, though he didn't get a salary.

guns, no cannons, and no military training. George Washington knew that he had an almost impossible job. He said to Patrick Henry, "Remember, Mr. Henry, what I now tell you: from the day I enter upon the command of the American armies, I date my fall, and the ruin of my reputation."

And that, too, is part of what made Washington great. He was willing to do what he thought was right and important even if it might bring his own ruin. (Of course, we know it didn't bring his ruin. It made him famous for all time.)

George Washington set out for Boston to take charge of the soldiers gathered there. In the meantime, the Continental Congress tried once more to patch things up with England. They sent another petition to King George III. This one was called the Olive Branch Petition. An olive branch is a symbol of peace. The colonists asked the king to consider their problems. But George wouldn't even read the petition.

Now all this may seem strange. The colonists were petitioning England and at the same time they were getting ready to fight. But most members of the Congress weren't ready to break away from England. Those who were—like Washington and Adams and Jefferson—were wise enough not to rush the others. People thought of separation from England as different from revolution. They wanted a revolution. That word had a splendid sound to it. Everyone knew of the Glorious Revolution of 1688. Englishmen and women were proud of that peaceful revolution and of the rights it gave them.

George III would not read the Olive Branch Petition (below), which was the colonists' last-ditch attempt to get him to think about their problems.

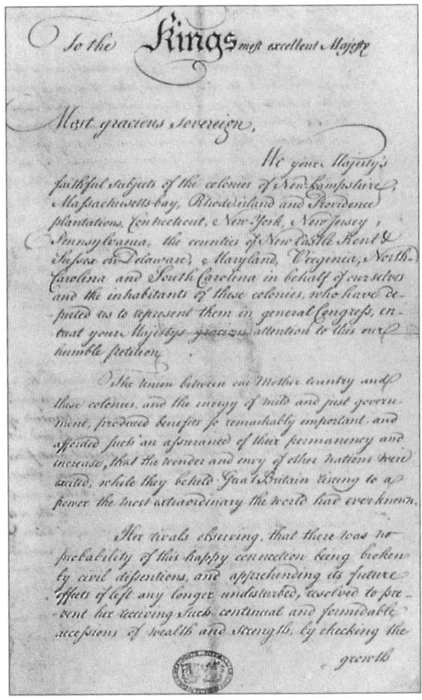

To the Kings most excellent Majesty

Most gracious Sovereign,

We your Majesty's faithful subjects of the colonies of New-Hampshire, Massachusetts-bay, Rhode-island and Providence plantations, Connecticut, New-York, New-Jersey, Pennsylvania, the counties of New Castle Kent & Sussex on Delaware, Maryland, Virginia, North-Carolina and South Carolina in behalf of ourselves and the inhabitants of these colonies, who have deputed us to represent them in general Congress, entreat your Majestys gracious attention to this our humble petition.

The union between our Mother country and these colonies, and the energy of mild and just government, produced benefits so remarkably important and afforded such an assurance of their permanency and increase, that the wonder and envy of other nations were excited, while they beheld Great Britain rising to a power the most extraordinary the world had ever known.

Her rivals observing, that there was no probability of this happy connection being broken by civil dissentions, and apprehending its future effects if left any longer undisturbed, resolved to prevent her receiving such continual and formidable accessions of wealth and strength, by checking the

growth

The Second Continental Congress is best known for two things:

1. Naming George Washington as general of the American armies.

2. Producing the Declaration of Independence. (It took more than a year to get that done.)

The Congress did more than that, but those two accomplishments were enough to make any body famous. (Yes, a congress is a body—a legislative body.)

For a long time many Americans thought they could have the rights of free people and still be part of the British empire. (And they might have, if the British king and Parliament had been wiser.)

Here's something about the American Revolution that not many Americans know. Some English citizens were rooting for the Americans. They knew that George was not a good king, and they didn't like his ministers either. They realized that some of their own precious English rights were being threatened because the king wanted more power for himself. As it turned out, the American Revolution helped bring better government to England.

The ideas that came out of our revolution soon infected the whole world. Monarchs and despots everywhere began trembling over those ideas of freedom and equality. Some kings and queens would lose their jobs because of those ideas. In France they would lose their heads. But that's another story—and a good one, too—that you'll have to read on your own.

This book is about America. People here were getting angry and saying and doing wild things. Soon there would be no turning back.

A Society of Patriotic Ladies

A British cartoon sneered at the North Carolina ladies.

If you ever happen to visit Edenton, North Carolina, you may see a big bronze teapot. It marks the place where Elizabeth King's house stood and where, in 1774, 51 women had a political meeting and agreed not to drink English tea. They said they would brew raspberry leaves for tea, and that they would also stop using English fabrics to make their clothes.

They weren't the only women drinking home brews. In Williamsburg, a dame dipped her goose quill into an inkpot and penned these lines:

Farewell to the Tea Board, with its gaudy Equipage,
Of Cups and Saucers, Cream Bucket, Sugar Tongs,

Then she went on with verses about how much she would miss drinking tea and gossiping with her friends around the tea table; but it was worth it, she concluded, because:

LIBERTY'S the Goddess I would choose
To reign triumphant in America.

18 The War of the Hills

A Pennsylvania infantryman in a spiffy uniform (looking handsome for the portrait painter).

England's Major John Pitcairn to the Earl of Sandwich (Boston, March 4, 1775):

> *I am satisfied that one active campaign, a smart action, and burning two or three of their towns, will set everything to rights. Nothing now, I am afraid, but this will ever convince those foolish bad people that England is in earnest.*

And so there was war. It seemed to begin almost by itself. Some people—on both sides—wanted to fight, and that was enough.

Two days after the Second Continental Congress appointed George Washington commander in chief of the Continental army—before anyone in Boston even knew there was a general—redcoats and Patriots were killing each other. They were fighting the first major battle of the Revolutionary War.

Two hills, Breed's and Bunker, lie just across the Charles River from Boston. Like Boston itself, they are on a peninsula connected to the mainland by a narrow neck: the Charlestown peninsula.

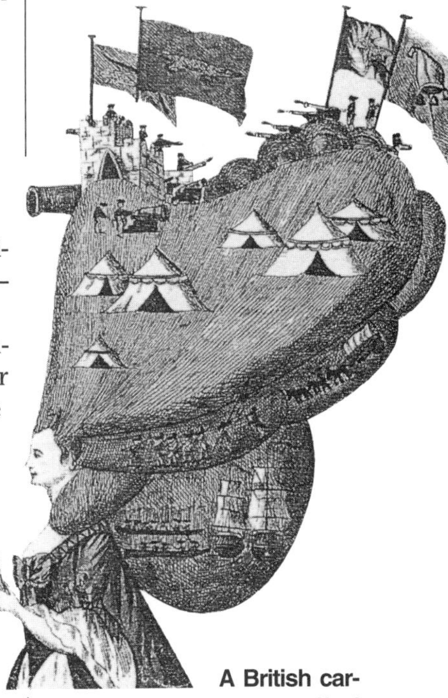
A British cartoon called "Noddle Island" (a real island outside Boston) made a double joke about women's fashions and the mistakes of Bunker Hill. (*Noddle* is an old English word for head.)

The British were asleep on that June night in 1775 when the Massachusetts soldiers began to dig fortifications on Breed's Hill. The Americans worked all night. They must have worked with great speed and ability because by morning it was done. Those hills that looked out on Boston were filled with troops and trenches.

The British couldn't believe it. For months they had tried to get the

Right: British naval forces in Boston harbor fire on Charlestown to back up the troops attacking the Patriots on foot.

Dr. Joseph Warren died on Breed's Hill. A Loyalist onlooker had this to say about him: "Since Adams went to Philadelphia, one Warren, a rascally patriot and apothecary of this town, has had the lead in the Provincial Congress....This fellow was happily killed, in coming out of the trenches the other day....You may judge what the herd must be like when such a one is their leader."

colonists to work for them. They needed barracks, and there were other construction jobs to be done. But nothing got finished. The Americans are lazy, the British thought. And then they saw this amazing feat, accomplished overnight. Breed's Hill was swarming with men and covered with impressive earthworks. Bunker Hill was dark with men. The British—especially the four British generals in Boston—were dumbfounded.

If they had thought a minute, they might have sent troops to capture the neck of the Charlestown peninsula—and perhaps trap the colonial soldiers. But they didn't think. They reacted.

Before long, barges filled with English soldiers were splashing their way from Boston, across the Charles River to Charlestown. Fifers played, drums pounded, and cannon blasted.

The British troops made ready to attack—head on. The Massachusetts men, dug in at the top of the hill, must have been scared—really scared. They had no training for this; they were fighting Europe's best soldiers; and they had very little gunpowder. They knew they had to use that gunpowder carefully. They had few bayonets; the British soldiers all had bayonets. The American officers told the volunteer soldiers to wait until the British soldiers were almost on top of them before they fired. "Wait until you see the whites of their eyes," they said.

And that is what the Massachusetts men did. Can you imagine the strain? It is said that those who saw the Battle of Bunker Hill never forgot the sounds, the smells, the ferocity, and the fear of that day. Pretend you are up there with them on the top of Breed's Hill. Watch the redcoats advance toward you, bayonets pointed. Don't panic, and don't fire until you hear the order to do so.

The story continues on page 93

Nose Blowing

"Colo. Washington of this Colony [Virginia] being appointed Generalissimo of all the American forces raised...made a demand of 500 Riflemen from the frontiers," wrote schoolmaster John Harrower in a letter to his family in Scotland. "So many men volunteered that Washington had to find a way to choose between them. He took a board of a foot square and with Chalk drew the shape of a moderate nose in the center and nailed it up to a tree at 150 yds. distance and those who came nighest [nearest] the mark with a single ball was to go. But by the first 40 or 50 the nose was all blown out of the board."

Guns and Swords

A bayonet is a sharp sword attached to a musket. At the time of the Revolutionary War most rifles did not have bayonets attached. Muskets and rifles look much alike, but the grooving inside the barrel of a rifle makes the bullet fly straight as aimed. The Revolutionary War generals on both sides expected their men to use muskets and bayonets. Reloading a gun took time; the bayonet was ready and deadly, and could be used at close quarters. The American farmers and frontiersmen brought their own rifles to war and understood their value before the military officers did. It took a long time for the officers to catch on. Muskets were still being used in the Civil War, nearly a century later.

John Harrower also said in his letter that many British officers had been killed "owing to the Americans taking sight when they fire." Taking sight when they fire? Yes—instead of just shooting, they were aiming their rifles and hitting their targets. That was new in warfare.

The Little Drummer Boy

Robert Steel, a drummer boy in Ephraim Doolittle's regiment from Cambridge, told of his part in the battle:

I beat to "Yankee Doodle" when we mustered for Bunker Hill that morning...the British ...marched with rather a slow step nearly up to our entrenchment, and the battle began. The conflict was sharp, but the British soon retreated with a quicker step than they came up, leaving some of their killed and wounded in sight of us....I was standing by the side of Benjamin Ballard, a Boston boy about my age...when one of our sergeants came up to us and said, "You are young and spry, run in a moment to some of the stores and bring some rum. Major Moore is badly wounded."...We went into a store, but see no one there....a man answered us from the cellar below....I asked him why he stayed down in the cellar. He answered, "To keep out of the way of the shot," and then said, "...take what you please."...I seized a brown, two quart earthen pitcher and filled [it] with rum...Ben took a pail and filled with water, and we hastened back to the entrenchment on the hill...Our rum and water went very quick. It was very hot, but I saved my pitcher and kept it for some time afterwards.

It was eerie, they say. All those soldiers climbing and no one firing.

Then, all at once, the hills seemed to explode. Bullets tore through the red coats and left the ground covered with bodies and blood. The British would not consider defeat or retreat. They landed more troops, and again the American fighters held their fire until it could hurt the most. The English soldiers kept coming, and falling, until "some had only eight or nine men a company left; some only three, four or five."

Suddenly it was quiet. This time the British made it to the top of the hill. The Americans were gone. They had run out of gunpowder. The British captured Breed's Hill and Bunker Hill, too. But what a price for two unimportant hills! More than 1,000 British soldiers were killed or wounded that day. The Americans lost 441 men.

Dr. Joseph Warren was one of those who died. He was a leader of the Boston Patriots. They say he was cool and brave under fire and that he inspired those around him. The same kinds of things are said of the handsome Major John Pitcairn, who fought for the Royal British Marines at Bunker Hill and didn't live to tell of it.

Left, an artist's impression of Bunker Hill: "a battle that should never have been fought on a hill that should never have been defended."

Firing a Revolutionary Cannon

Firing a Revolutionary War cannon isn't easy; six or seven men are needed to do the job. And it is dangerous: sometimes the monsters explode. Firing begins when the officer in charge shouts, *"Worm!"* A wormer—a soldier with a long, corkscrew-shaped iron worm—twists the worm and cleans out the barrel. Next comes the call *"Sponge!"* and a sponger sticks a wet sheepskin into the gun barrel. That cools it down and puts out sparks. *"Load!"* says the officer, and a bag of powder is stuffed into the barrel, followed by a big iron ball, or grapeshot (clusters of small balls that scatter with great force, killing or wounding men over a broad area). *"Ram!"* Now a rammer, holding a pole with a wooden disk on its end, pushes and packs the ammunition. *"Pick and prime!"* A gunner sticks a pick into the barrel and breaks open the ammunition sack. He adds powder in a vent hole, and puts a pinch of powder on top of the cannon barrel. *"Give!"* shouts the officer, and the gunner lights a slow fuse.

"Fire!" The gunner uses the fuse to light the powder on top of the barrel. The flame skips through the vent and sets off the powder inside the cannon. The ball explodes out of the gun's mouth at a speed of about 1,000 feet a second. *Watch out!*

19 Fighting Palm Trees

Why Did I Go?

"Captain Preston, why did you go to the Concord fight, the 19th April, 1775?" Judge Mellen Chamberlain asked old Captain Levi Preston, years after the battle.

"Why did I go?" repeated Preston.

"Yes, my histories tell me that you men of the Revolution took up arms against intolerable oppression."

"What were they? Oppressions? I didn't feel them."

"What," said the judge, "were you not oppressed by the Stamp Act?"

L.P.: I never saw one of those stamps...

M.C.: Well, what then about the tea tax?

L.P.: Tea tax! I never drank a drop of the stuff; the boys threw it all overboard...

M.C.: Well, then, what was the matter? and what did you mean in going to fight?

L.P.: Young man, what we meant in going for those redcoats was this: we had always governed ourselves, and we always meant to. They didn't mean we should.

Lord North didn't want to run a war. "On military matters," he said, "I speak ignorantly and therefore without effect."

"What is all this fuss about a little tax on tea?" said some people in England. "Those American colonists are an ungrateful bunch," said others. "Punish them! Show them Britain's power!" said still others.

Englishmen and women argued about what to do with the colonies. William Pitt said, "You cannot but respect their cause." Pitt said it was the spirit of liberty that was making the colonists protest against British taxes. It was "the same spirit which established...that no subject of England shall be taxed but by his own consent." Pitt, who had practically run the British government back during the French and Indian War, was retired and ill. But he was still a powerful speaker and a friend of America.

King George didn't much like Pitt. It was Lord North who now held power, and Lord North believed the colonies should be taught a lesson. Most English people and most members of Parliament seemed to agree with Lord North.

A Mr. Van stood up in the House of Commons (Parliament has two houses, like our Congress) and said that he was "of the opinion that the town of Boston ought to be knocked about their ears and destroyed." Then he continued, "you will never meet with that proper obedience to the laws of this country until you have destroyed that nest of hornets."

But Boston was a hornet's nest that wasn't easy to destroy. "What about Charleston?" King George's ministers asked. They had heard that some Liberty Boys in Charleston were gathering under a tree and making trouble. A few shots from British cannons and they would run, said the king's men. The mighty British navy would scare those upstart American provincials in South Carolina, they added. And so a fleet of armed ships and seven regiments of soldiers were sent across the ocean.

An unfinished fort stood on Sullivan's Island in Charleston's harbor. It had double walls of palmetto logs placed 16 feet apart. Sand was packed between the palmetto walls. But only the front of the fort was completed, the sides were half done, and the back was open. General Charles Lee, who had been sent south by General Washington, took one look and called it a "slaughter pen." He suggested that the fort be abandoned.

South Carolina's governor, John Rutledge (whom some people were calling Dictator John because he always seemed to get his way), insisted that the fort be defended. Colonel William Moultrie, who was placed in command of the fort, believed he could do it.

The British ships sailed grandly into the harbor—and ran aground. That means that some of them got stuck on shoals (which are sandbars). Their ships' pilots didn't know the harbor and its safe passageways. Since they were stuck anyway, they decided they might as well destroy the fort, unload their men, and take Sullivan's Island.

And so they blasted their cannons—and then something unbelievable happened. Their shells stuck in the sides of the fort. The soft palmetto wood, and the thick sand walls, absorbed the shells as a sponge might. The walls just held on to the cannonballs. The British naval experts had never seen anything like this. And the soldiers who were supposed to march onto the island? Well, the British had been misinformed about the depth of the water. It was too deep, and the men couldn't get to the island. It was "unspeakable mortification," said a British general. And what about those ships, stuck on the shoals? What kind of targets did they make? You guessed it—perfect targets.

The redcoats, led by Sir Henry Clinton (right), failed to take Fort Sullivan, defended by Colonel Moultrie (below).

Charleston harbor (above) was full of sandbanks, which many of the British boats got stranded on when they tried to sail in.

Mortification is distressing humiliation.

Now do you know why South Carolina's flag has a palmetto tree on it?

Moultrie's guns turned the British ships into a slaughterhouse. The flagship *Bristol* was hit 70 times.

To get on with this story—when the British finally limped out of Charleston harbor, not one of their ships was undamaged. Some were destroyed. Here is a poem describing the battle, said to be written by Sir Peter Parker, who was in charge of the British fleet at Charleston. See if you think Sir Peter actually wrote it. (Before you begin, you need to know that Falstaff and Pistol were comic characters—who talked tougher than they acted—from plays by William Shakespeare. The *Bristol* was the name of one of Sir Peter's ships.)

My Lords, with your leave
An account I will give
That deserves to be written in meter;
For the rebels and I
Have been pretty nigh—
Faith! almost too nigh for Sir Peter.

A London **journal, the *Annual Register*, commented on the disaster at Sullivan's Island: "To suppose that the Generals...should have been 19 days in that small island, without ever examining until the very instant of action the nature of the only passage by which they could render service to their friends and fellows, fulfill the purpose of their landing and answer the ends for which they were embarked in the expedition would seem a great defect in military prudence and circumspection."**

Which was a long way of saying that the British forces in Charleston didn't have a clue about what they were doing.

With much labor and toil
Unto Sullivan's Isle
I came fierce as Falstaff or Pistol,
But the Yankees (God rot 'em)—
I could not get at 'em—
Most terribly mauled my poor Bristol.

Devil take 'em; their shot
Came so swift and so hot,
And the cowardly dogs stood so stiff, sirs,
That I put ship about
And was glad to get out,
Or they would not have left me a skiff, sirs!

But, my lords, do not fear
For before the next year,
(Altho' a small island could fret us),
The Continent whole
We shall take, by my soul,
If the cowardly Yankee will let us.

When England's forces again came South, they took their "cowardly" foe more seriously.

One of the oddest American generals (he was more interested in dogs than people), Charles Lee argued against defending Sullivan's Island (below, with detailed plans of the fort and ships' positions.)

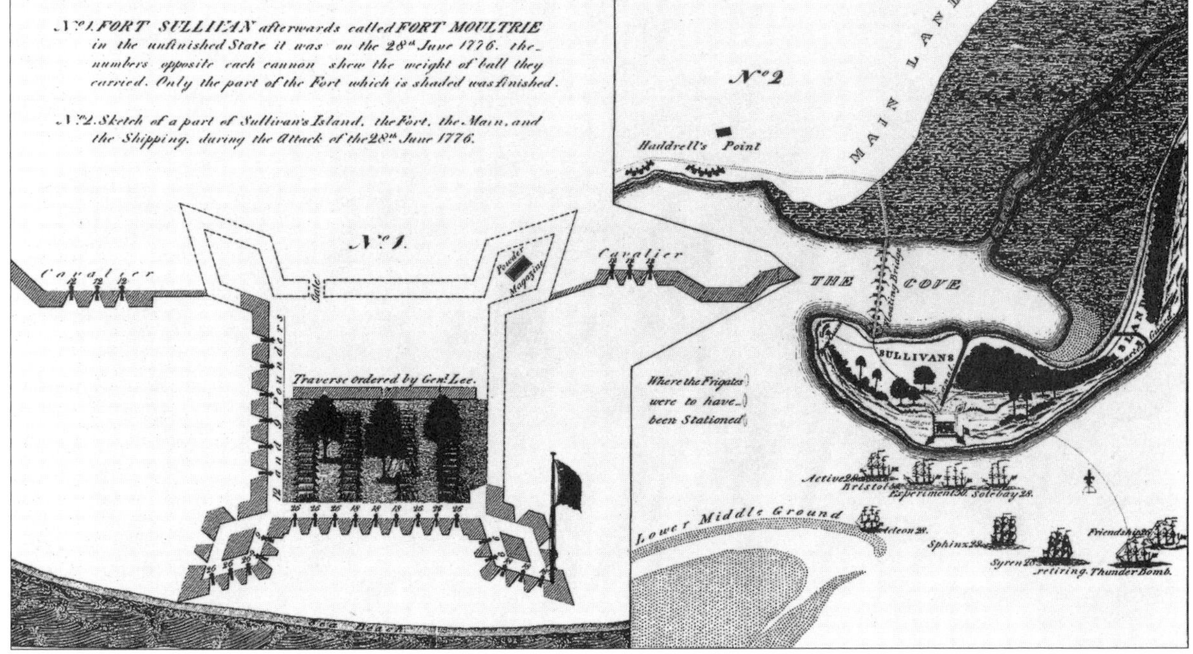

20 Declaring Independence

John Adams said that Jefferson should write the Declaration. "Well," said Jefferson, "if you are decided, I will do as well as I can."

Unless you like to memorize dates, there aren't many that you need to remember. But here are a few that are important:

1215
1492
1607
1620

What happened in those years? You can't remember? Go ahead and try. You might be surprised and find there is more in your brain than you realize. (Then, if you need to check, look at the end of this chapter.)

Now I have another date for you to remember, and this one is the most important of all. Something happened on that day that changed America—it even changed the whole world. (It was a day that King George III didn't think important. He would find out how wrong he was.)

The date is July 4, 1776. That was the day the members of the Second Continental Congress approved a Declaration of Independence. It was a year after the Battle of Bunker Hill, and, finally, the Americans had made up their minds to be free of Great Britain.

But that wasn't why the world was changed. It was the words they used in that declaration that made all the difference.

The delegates believed that if they were going to vote for independence, they should have a good reason. They knew that when they signed the declaration they became traitors to England. They would each be hanged if England captured them.

Thomas Jefferson thought John Adams should write the document about independence. This is how Adams remembered their conversation:

"Why will you not?" Jefferson asked. "You ought to do it."

"Reasons enough."

"What can be your reasons?"

"Reason 1st. You are a Virginian, and a Virginian ought to appear at the head of this business. Reason 2d. I am obnoxious, suspected, and unpopular. You are very much otherwise. Reason 3d. You can write ten times better than I can."

If they were going to take that big risk, they wanted to make it worthwhile. And it would be worthwhile if they could help create a free nation, a great nation, a nation run by its citizens—something that had never before been done.

So they thought it important to explain exactly what they were doing and why it was necessary to be free of English rule.

That's why they asked Thomas Jefferson, one of the members of the congress, to write a paper—called a declaration—that would:

- ***tell their beliefs about good government,***
- ***tell what King George had done wrong, and***
- ***announce that the colonies were now free and independent states.***

Some people thought it surprising that Thomas Jefferson was asked to write the declaration. Jefferson was one of the youngest members of the Continental Congress. He was a tall, shy redhead who loved to read, run, ride horseback, and play the violin. He had a reputation for writing well. John Adams said of him, "Though a silent member in Congress, he was so prompt, frank...and decisive upon committees and in conversation—not even Samuel Adams was more so—that he soon seized upon my heart."

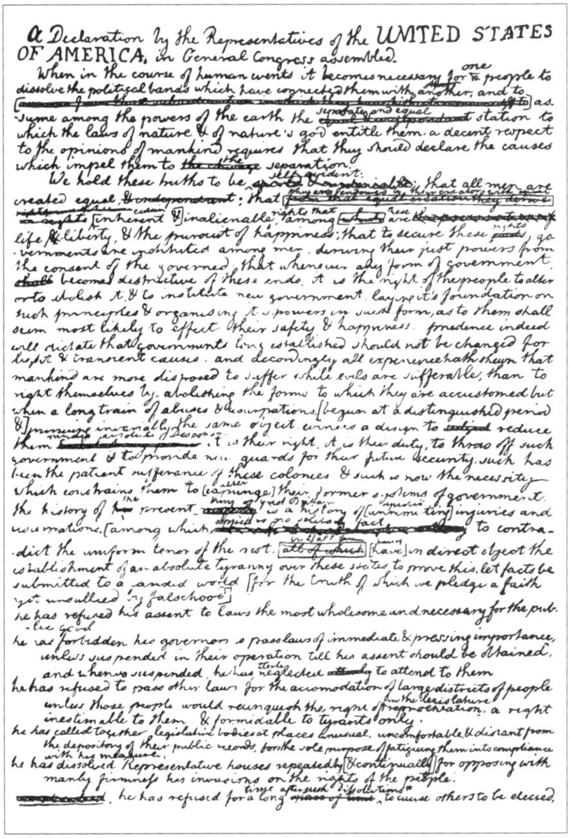
A Declaration by the Representatives of the UNITED STATES OF AMERICA, in General Congress assembled.

When in the course of human events it becomes necessary for one people to dissolve the political bands which have connected them with another, and to assume among the powers of the earth the separate and equal station to which the laws of nature & of nature's god entitle them, a decent respect to the opinions of mankind requires that they should declare the causes which impel them to the separation.

We hold these truths to be self-evident; that all men are created equal; that they are endowed by their creator with equal rights [inherent &] inalienable, among these are life & liberty, & the pursuit of happiness; that to secure these rights, governments are instituted among men, deriving their just powers from the consent of the governed; that whenever any form of government becomes destructive of these ends, it is the right of the people to alter or to abolish it, & to institute new government, laying it's foundation on such principles & organising it's powers in such form, as to them shall seem most likely to effect their safety & happiness. prudence indeed will dictate that governments long established should not be changed for light & transient causes. and accordingly all experience hath shewn that mankind are more disposed to suffer while evils are sufferable, than to right themselves by abolishing the forms to which they are accustomed but when a long train of abuses & usurpations [begun at a distinguished period &] pursuing invariably the same object, evinces a design to reduce them under absolute Despotism, it is their right, it is their duty, to throw off such government & to provide new guards for their future security. such has been the patient sufferance of these colonies; & such is now the necessity which constrains them to [expunge] their former systems of government. the history of the present king of Great Britain is a history of [unremitting] injuries and usurpations, [among which appears no solitary fact to contradict the uniform tenor of the rest, [have] in direct object the establishment of an absolute tyranny over these states. to prove this, let facts be submitted to a candid world [for the truth of which we pledge a faith yet unsullied by falsehood]

he has refused his assent to laws the most wholesome and necessary for the public good.

he has forbidden his governors to pass laws of immediate & pressing importance, unless suspended in their operation till his assent should be obtained; and when so suspended, he has utterly neglected to attend to them.

he has refused to pass other laws for the accomodation of large districts of people unless those people would relinquish the right of representation in the legislature, a right inestimable to them, & formidable to tyrants only.

he has called together legislative bodies at places unusual, uncomfortable & distant from the depository of their public records, for the sole purpose of fatiguing them into compliance with his measures.

he has dissolved Representative houses repeatedly [& continually] for opposing with manly firmness his invasions on the rights of the people.

he has refused for a long time after such dissolutions to cause others to be elected

The first draft of the Declaration of Independence, in Jefferson's handwriting.

Jefferson had left his wife behind on his Virginia farm, and he missed her. He wasn't sure he could write a good declaration. But John Adams and Benjamin Franklin had faith in him. They talked Thomas Jefferson into trying. Adams told him, "You can write ten times better than I can."

Adams and Franklin were right. Thomas Jefferson knew just what to say, and he said it in a way that inspired people all over the world.

The whole declaration is something to read and think about, but one part will ring in your ears with its greatness. Jefferson wrote:

> *We hold these truths to be self-evident, that all men are created equal, that they are endowed by their Creator with certain unalienable Rights, that among these are Life, Liberty and the pursuit of Happiness. That to secure these rights, Governments are instituted among Men, deriving their just powers from the consent of the governed.*

That was plain language in the 18th century, but you might have to

Resistance to tyrants is obedience to God.

—MOTTO ON THOMAS JEFFERSON'S SEAL, CIRCA 1776

King George, in an English cartoon, as the helpless rider of "The horse America, throwing his Master."

read it a few times to understand it. It is worth doing. Those words are worth memorizing.

All men are created equal.

Just what does "equal" mean?

Are we all the same? Look around you. Of course we aren't. Some of us are smarter than others, and some of us are better athletes, and some of us are better looking, and some are nicer. But none of that matters, said Jefferson. We are all equal in the eyes of God, and we are all entitled to equal rights: the right to live, the right to be free, the right to be able to try to find the kind of life that will make us happy.

And that is the whole reason for having governments, he said. Governments are not made to make kings happy. They are for the benefit of the people who are being governed. Governments should have "the consent of the governed."

Sometimes, when ideas are written down, they take on meanings that go beyond what the writers intended. Jefferson's Declaration of Independence was great from the moment he wrote it, but it has grown even greater with the passing of time. He said "all men are created equal." He didn't mention women. Did he mean to include women? No one knows. Perhaps not. We do know that in the 18th century the words "men" and "mankind" included men and women. But very few people, except for Tom Paine, thought much about women's

Dates to Remember—or Else!

1215: Magna Carta
1492: Columbus sails to America
1607: Jamestown settled
1620: the Pilgrims land at Plymouth
1776: the Declaration of Independence!

Here's another date for your memory bank:
1610: when Spanish speakers founded Santa Fe, New Mexico.

rights. It was the 20th century before women in America had the right to vote.

Did Thomas Jefferson mean to include black men when he said "all men"? Historians sometimes argue about that. You'll have to decide for yourself.

In 1776, when Jefferson wrote the Declaration, he included a long section in which he described slavery as a "cruel war against human nature." Yet Jefferson lived in a slave society and owned slaves himself.

He thought slavery was wrong, and he said so. "Nothing is more certainly written in the book of fate than that these people are to be free," wrote Jefferson. Many congressmen agreed. John Adams spoke out strongly against slavery. Benjamin Franklin and Benjamin Rush founded the first antislavery society in the New World. But South Carolina and Georgia would not sign the Declaration if it contained the antislavery section. So Jefferson's antislavery words were taken out. The delegates compromised.

Should they have gone ahead without those southern colonies? That would have meant that the Deep South would not have joined in the fight against England. It might have meant defeat for the proposed union of states.

The appointment of a woman to office is an innovation for which the public is not prepared, nor am I.
—THOMAS JEFFERSON TO ALBERT GALLATIN, 1807

Nature has given to our black brethren talents equal to those of the other colours of men.
—THOMAS JEFFERSON TO BENJAMIN BANNEKER, SLAVE-BORN INVENTOR, 1792

Jefferson and Adams and Franklin and others thought the Union was more important than the issue of slavery. They knew that staying with England would not bring freedom to the slaves. They thought slavery could be dealt with later. Do you agree with them?

Those were tough decisions the delegates were making.

It took a civil war to end slavery. Do you think that war could have been avoided? Do you think the delegates should have acted differently in 1776?

Of one thing you can be sure. Today, when people all over the world read Jefferson's words, they understand them to mean all people—men, women, and children—of all colors and beliefs.

In 1776 Arthur Middleton (left), of South Carolina, owned more than 50,000 acres and 800 slaves. He also signed the Declaration of Independence.

The Declaration of Independence

IN CONGRESS, JULY 4, 1776.

A DECLARATION

BY THE REPRESENTATIVES OF THE

UNITED STATES OF AMERICA,

IN GENERAL CONGRESS ASSEMBLED.

WHEN in the Courſe of human Events, it becomes neceſſary for one People to diſſolve the Political Bands which have connected them with another, and to aſſume among the Powers of the Earth, the ſeparate and equal Station to which the Laws of Nature and of Nature's God entitle them, a decent Reſpect to the Opinions of Mankind requires that they ſhould declare the cauſes which impel them to the Separation.

We hold theſe Truths to be ſelf-evident, that all Men are created equal, that they are endowed by their Creator with certain unalienable Rights, that among theſe are Life, Liberty, and the Purſuit of Happineſs—That to ſecure theſe Rights, Governments are inſtituted among Men, deriving their juſt Powers from the Conſent of the Governed, that whenever any Form of Government becomes deſtructive of theſe Ends, it is the Right of the People to alter or to aboliſh it, and to inſtitute new Government, laying its Foundation on ſuch Principles, and organizing its Powers in ſuch Form, as to them ſhall ſeem moſt likely to effect their Safety and Happineſs. Prudence, indeed, will dictate that Governments long eſtabliſhed ſhould not be changed for light and tranſient Cauſes; and accordingly all Experience hath ſhewn, that Mankind are more diſpoſed to ſuffer, while Evils are ſufferable, than to right themſelves by aboliſhing the Forms to which they are accuſtomed. But when a long Train of Abuſes and Uſurpations, purſuing invariably the ſame Object, evinces a Deſign to reduce them under abſolute Deſpotiſm, it is their Right, it is their Duty, to throw off ſuch Government, and to provide new Guards for their future Security. Such has been the patient Sufferance of theſe Colonies; and ſuch is now the Neceſſity which conſtrains them to alter their former Syſtems of Government. The Hiſtory of the preſent King of Great-Britain is a Hiſtory of repeated Injuries and Uſurpations, all having in direct Object the Eſtabliſhment of an abſolute Tyranny over theſe States. To prove this, let Facts be ſubmitted to a candid World.

He has refuſed his Aſſent to Laws, the moſt wholeſome and neceſſary for the public Good.

He has forbidden his Governors to paſs Laws of immediate and preſſing Importance, unleſs ſuſpended in their Operation till his Aſſent ſhould be obtained; and when ſo ſuſpended, he has utterly neglected to attend to them.

He has refuſed to paſs other Laws for the Accommodation of large Diſtricts of People, unleſs thoſe People would relinquiſh the Right of Repreſentation in the Legiſlature, a Right ineſtimable to them, and formidable to Tyrants only.

He has called together Legiſlative Bodies at Places unuſual, uncomfortable, and diſtant from the Depoſitory of their public Records, for the ſole Purpoſe of fatiguing them into Compliance with his Meaſures.

He has diſſolved Repreſentative Houſes repeatedly, for oppoſing with manly Firmneſs his Invaſions on the Rights of the People.

He has refuſed for a long Time, after ſuch Diſſolutions, to cauſe others to be elected; whereby the Legiſlative Powers, incapable of Annihilation, have returned to the People at large for their exerciſe; the State remaining in the mean time expoſed to all the Dangers of Invaſion from without, and Convulſions within.

He has endeavoured to prevent the Population of theſe States; for that Purpoſe obſtructing the Laws for Naturalization of Foreigners; refuſing to paſs others to encourage their Migrations hither, and raiſing the Conditions of new Appropriations of Lands.

He has obſtructed the Adminiſtration of Juſtice, by refuſing his Aſſent to Laws for eſtabliſhing Judiciary Powers.

He has made Judges dependent on his Will alone, for the Tenure of their Offices, and the Amount and Payment of their Salaries.

He has erected a Multitude of new Offices, and ſent hither Swarms of Officers to harraſs our People, and eat out their Subſtance.

He has kept among us, in Times of Peace, Standing Armies, without the conſent of our Legiſlatures.

He has affected to render the Military independent of and ſuperior to the Civil Power.

He has combined with others to ſubject us to a Juriſdiction foreign to our Conſtitution, and unacknowledged by our Laws; giving his Aſſent to their Acts of pretended Legiſlation:

For quartering large Bodies of Armed Troops among us:

For protecting them, by a mock Trial, from Puniſhment for any Murders which they ſhould commit on the Inhabitants of theſe States:

For cutting off our Trade with all Parts of the World:

For impoſing Taxes on us without our Conſent:

For depriving us, in many Caſes, of the Benefits of Trial by Jury:

For tranſporting us beyond Seas to be tried for pretended Offences:

For aboliſhing the free Syſtem of Engliſh Laws in a neighbouring Province, eſtabliſhing therein an arbitrary Government, and enlarging its Boundaries, ſo as to render it at once an Example and fit Inſtrument for introducing the ſame abſolute Rule into theſe Colonies:

For taking away our Charters, aboliſhing our moſt valuable Laws, and altering fundamentally the Forms of our Governments:

For ſuſpending our own Legiſlatures, and declaring themſelves inveſted with Power to legiſlate for us in all Caſes whatſoever.

He has abdicated Government here, by declaring us out of his Protection and waging War againſt us.

He has plundered our Seas, ravaged our Coaſts, burnt our Towns, and deſtroyed the Lives of our People.

He is, at this Time, tranſporting large Armies of foreign Mercenaries to compleat the Works of Death, Deſolation, and Tyranny, already begun with circumſtances of Cruelty and Perfidy, ſcarcely paralleled in the moſt barbarous Ages, and totally unworthy the Head of a civilized Nation.

He has conſtrained our fellow Citizens taken Captive on the high Seas to bear Arms againſt their Country, to become the Executioners of their Friends and Brethren, or to fall themſelves by their Hands.

He has excited domeſtic Inſurrections amongſt us, and has endeavoured to bring on the Inhabitants of our Frontiers, the merciless Indian Savages, whoſe known Rule of Warfare, is an undiſtinguiſhed Deſtruction, of all Ages, Sexes and Conditions.

In every ſtage of theſe Oppreſſions we have Petitioned for Redreſs in the moſt humble Terms: Our repeated Petitions have been anſwered only by repeated Injury. A Prince, whoſe Character is thus marked by every act which may define a Tyrant, is unfit to be the Ruler of a free People.

Nor have we been wanting in Attentions to our Brittiſh Brethren. We have warned them from Time to Time of Attempts by their Legiſlature to extend an unwarrantable Juriſdiction over us. We have reminded them of the Circumſtances of our Emigration and Settlement here. We have appealed to their native Juſtice and Magnanimity, and we have conjured them by the Ties of our common Kindred to diſavow theſe Uſurpations, which, would inevitably interrupt our Connections and Correſpondence. They too have been deaf to the Voice of Juſtice and of Conſanguinity. We muſt, therefore, acquieſce in the Neceſſity, which denounces our Separation, and hold them, as we hold the reſt of Mankind, Enemies in War, in Peace, Friends.

We, therefore, the Repreſentatives of the UNITED STATES OF AMERICA, in GENERAL CONGRESS, Aſſembled, appealing to the Supreme Judge of the World for the Rectitude of our Intentions, do, in the Name, and by Authority of the good People of theſe Colonies, ſolemnly Publiſh and Declare, That theſe United Colonies are, and of Right ought to be, FREE AND INDEPENDENT STATES; that they are abſolved from all Allegiance to the Britiſh Crown, and that all political Connection between them and the State of Great-Britain, is and ought to be totally diſſolved; and that as FREE AND INDEPENDENT STATES, they have full Power to levy War, conclude Peace, contract Alliances, eſtabliſh Commerce, and to do all other Acts and Things which INDEPENDENT STATES may of right do. And for the ſupport of this Declaration, with a firm Reliance on the Protection of divine Providence, we mutually pledge to each other our Lives, our Fortunes, and our ſacred Honor.

Signed by ORDER *and in* BEHALF *of the* CONGRESS,

JOHN HANCOCK, PRESIDENT.

We hold these Truths to be self-evident, that all Men are created equal, that they are endowed by their Creator with certain unalienable Rights, that among these are Life, Liberty, and the Pursuit of Happiness—That to secure these Rights, Governments are instituted among Men, deriving their just Powers from the Consent of the Governed, that whenever any Form of Government becomes destructive of these Ends, it is the Right of the People to alter or to abolish it, and to institute new Government, laying its Foundation on such Principles, and organizing its Powers in such Form, as to them shall seem most likely to effect their Safety and Happiness. Prudence, indeed, will dictate that Governments long established should not be changed for light and transient Causes; and accordingly all Experience hath shewn, that Mankind are more disposed to suffer, while Evils are sufferable, than to right themselves by abolishing the Forms to which they are accustomed. But when a long Train of Abuses and Usurpations, pursuing invariably the same Object, evinces a Design to reduce them under absolute Despotism, it is their Right, it is their Duty, to throw off such Government, and to provide new Guards for their future Security. Such has been the patient Sufferance of these Colonies; and such is now the Necessity which constrains them to alter their former Systems of Government. The History of the present King of Great-Britain is a History of repeated Injuries and

When in the Course of human Events, it becomes necessary for one People to dissolve the Political Bands which have connected them with another, and to assume among the Powers of the Earth, the separate and equal Station to which the Laws of Nature and of Nature's God entitle them, a decent Respect to the Opinions of Mankind requires that they should declare the causes which impel them to the Separation.

Usurpations, all having in direct Object the Establishment of an absolute Tyranny over these States. To prove this, let Facts be submitted to a candid World.

He has refused his Assent to Laws, the most wholesome and necessary for the public Good.

He has forbidden his Government to pass Laws of immediate and pressing Importance, unless suspended in their Operation till his Assent should be obtained; and when so suspended, he has utterly neglected to attend to them.

He has refused to pass other Laws for the Accommodation of large Districts of People, unless those People would relinquish the Right of Representation in the Legislature, a Right inestimable to them, and formidable to Tyrants only.

He has called together Legislative Bodies at Places unusual, uncomfortable, and distant from the Depository of their public Records, for the sole Purpose of fatiguing them into Compliance with his Measures.

He has dissolved Representative Houses repeatedly, for opposing with manly Firmness his Invasions on the Rights of the People.

He has refused for a long Time, after such Dissolutions, to cause others to be elected; whereby the Legislative Powers, incapable of Annihilation, have returned to the People at large for their exercise; the State remaining in the meantime exposed to all the Dangers of Invasion from without, and Convulsions within.

He has endeavoured to prevent the Population of these States; for that Purpose obstructing the Laws for Naturalization of Foreigners; refusing to pass others to encourage their Migrations hither, and raising the Conditions of new Appropriations of Lands.

He has obstructed the Administration of Justice, by refusing his Assent to Laws for establishing Judiciary Powers.

He has made Judges dependent on his Will alone, for the Tenure of their Offices, and the Amount and Payment of their Salaries.

He has erected a Multitude of new Offices, and sent hither Swarms of Officers to harrass our People, and eat out their Substance.

He has kept among us, in Times of Peace, Standing Armies, without the consent of our Legislatures.

He has affected to render the Military independent of and superior to the Civil Power.

He has combined with others to subject us to a Jurisdiction foreign to our Constitution, and unacknowledged by our Laws; giving his Assent to their Acts of pretended Legislation:

For quartering large Bodies of Armed Troops among us:

For protecting them, by a mock Trial, from Punishment for any Murders which they should commit on the Inhabitants of these States:

For cutting off our Trade with all Parts of the World:

For imposing Taxes on us without our Consent:

For depriving us, in many Cases, of the Benefits of Trial by Jury:

For transporting us beyond Seas to be tried for pretended Offences:

For abolishing the free System of English Laws in a neighbouring Province, establishing therein an arbitrary Government, and enlarging its Boundaries, so as to render it at once an Example and fit Instrument for introducing the same absolute Rule into these Colonies:

For taking away our Charters, abolishing our most valuable Laws, and altering fundamentally the Forms of our Governments:

For suspending our own Legislatures, and de-

The Declaration of Independence (continued)

claring themselves invested with Power to legislate for us in all Cases whatsoever.

He has abdicated Government here, by declaring us out of his Protection and waging War against us.

He has plundered our Seas, ravaged our Coasts, burnt our Towns, and destroyed the Lives of our People.

He is, at this Time, transporting large Armies of foreign Mercenaries to compleat the works of Death, Desolation, and Tyranny, already begun with circumstances of Cruelty and Perfidy, scarcely paralleled in the most barbarous Ages, and totally unworthy of the Head of a civilized Nation.

He has constrained our fellow Citizens taken Captive on the high Seas to bear Arms against their Country, to become the Executioners of their Friends and Brethren, or to fall themselves by their Hands.

He has excited domestic Insurrections amongst us, and has endeavoured to bring on the Inhabitants of our Frontiers, the merciless Indian Savages, whose known Rule of Warfare, is an undistinguished Destruction, of all Ages, Sexes and Conditions.

In every stage of these Oppressions we have Petitioned for Redress in the most humble Terms: Our repeated Petitions have been answered only by repeated Injury. A Prince, whose Character is thus marked by every act which may define a Tyrant, is unfit to be the Ruler of a Free People.

Nor have we been wanting in Attentions to our British Brethren. We have warned them from Time to Time of Attempts by their Legislature to extend an unwarrantable Jurisdiction over us. We have reminded them of the Circumstances of our Emigration and Settlement here. We have appealed to their native Justice and Magnanimity, and we have conjured them by the Ties of our common Kindred to disavow these Usurpations, which would inevitably interrupt our Connections and Correspondence. They too have been deaf to the Voice of Justice and of Consanguinity. We must, therefore, acquiesce in the Necessity, which denounces our Separation, and hold them, as we hold the rest of Mankind, Enemies in War, in Peace, Friends.

We, therefore, the Representatives of the UNITED STATES OF AMERICA, in GENERAL CONGRESS Assembled, appealing to the Supreme Judge of the World for the Rectitude of our Intentions, do, in the Name, and by the Authority of the good People of these Colonies, solemnly Publish and Declare, That these United Colonies are, and by Right ought to be, FREE AND INDEPENDENT STATES; that they are absolved from all Allegiance to the British Crown, and that all political Connection between them and the State of Great-Britain, is and ought to be totally dissolved; and that as FREE AND INDEPENDENT STATES, they have full Power to levy War, conclude Peace, contract Alliances, establish Commerce, and to do all other Acts and Things which INDEPENDENT STATES may of right do. And for the support of this Declaration, with a firm Reliance on the Protection of divine Providence, we mutually pledge to each other our Lives, our Fortunes, and our sacred Honor.

Signed by ORDER and in BEHALF of this CONGRESS,

John Hancock

JOHN HANCOCK, PRESIDENT.

21 Signing Up

Fifty-six men signed the Declaration of Independence. John Hancock (above) was the first.

It took courage to sign that Declaration. John Hancock was first to put his name down. He did it with a big, bold signature. "So the king doesn't have to put on his glasses," he is supposed to have said. (Because of that, today, when you sign a document, people sometimes call your signature a "John Hancock.")

John Dickinson of Pennsylvania wouldn't sign. He believed the Declaration was foolhardy. He thought the colonists should work to gain the rights of free citizens within the British Empire. "I had rather forfeit popularity forever, than vote away the blood and happiness of my countrymen," he said. Independence! To Dickinson that was "like destroying our house in winter...before we have got another shelter." But he loved America dearly, so after he refused to sign the Declaration of Independence, he enlisted in the Continental army as a private and fought for his country. And he was right; he did lose his popularity.

The citizens outside the red-brick Pennsylvania State House, where the delegates voted, were now screaming for independence. That didn't make it easy for the men inside. They knew they would pay with their lives if the colonial army was squashed by Britain. And all the power seemed on Great Britain's side.

The Pennsylvania State House was soon to be called Independence Hall. Why?

On July 4, 1776, the Declaration was officially approved by the delegates.

When the Declaration was read in New York, a mob pulled down the statue of George III in Bowling Green. It was made of 4,000 pounds of lead, from which it was estimated that 42,000 bullets could be cast.

After independence came, Adams went to England as ambassador, wrote a book on government that the Constitution makers read, and became the second president of the United States.

It was John Adams, perhaps more than anyone else, who got the delegates to sign the Declaration. Adams was a talker as well as a thinker. At the Second Continental Congress he kept talking and talking and talking until finally he convinced the delegates.

Then John Adams wrote to his wife, Abigail:

> *Yesterday, the greatest question was decided which ever was debated in America, and a greater, perhaps, never was nor will be decided among men. A Resolution was passed without one dissenting Colony, "that these United Colonies are, and of right ought to be, free and independent States...." You will see, in a few days, a Declaration setting forth the causes which have impelled us to this mighty revolution, and the reasons which will justify it in the sight of God and man.*

Standing by the table (left to right), John Adams, Roger Sherman, Robert Livingston, Thomas Jefferson, and Benjamin Franklin present the Declaration of Independence to Congress.

Copies of the Declaration were still warm from the printing press when they were put on coastal vessels or stuffed into saddlebags so they could be sped on their way to each of the 13 colonies. On July 9, the document reached New York and was read to General Washington's troops, who shouted hurrah and tossed their hats in the air. That night a gilded statue of George III on horseback was pulled down from its pedestal on Manhattan's Bowling Green. (The statue was soon melted down and turned into bullets.)

On July 19, the Declaration arrived in Boston, and Tom Crafts, a house painter, stepped out on a small square balcony in front of the Massachusetts State House and read it aloud. "When, in the course of human events," he began in his flat New England tone. When he finished a voice rang out, "God save the American States," and the crowd cheered mightily. Two days later Abigail Adams wrote a letter to John.

> *The bells rang, the privateers fired the forts and batteries, the cannon were discharged, the platoons followed, & every face appeared joyful. ...After dinner the King's [coat of] Arms were taken down from the State House & every vestige of him from every place in which it appeared, & burnt....Thus ends royal Authority in this State. And all the people shall say Amen.*

An Awful Silence

Benjamin Rush was one of the first doctors in America to have an interest in psychiatry, which is the study of illnesses of the mind. After the Revolution was over, Dr. Rush remembered the "fears and sorrows and sleepless nights" of those who signed the Declaration. Here is part of what he wrote in a letter to John Adams:

Dear Old Friend...Do you recall your memorable speech upon the day on which the vote was taken? Do you recall the pensive and awful silence which pervaded the house when we were called up, one after another, to the table of the President of Congress to subscribe what was believed by many at that time to be our own death warrants? The silence and gloom of the morning were interrupted, I well recollect, only for a moment by Colonel [Benjamin] Harrison of Virginia [who was heavy], who said to Mr. [Elbridge] Gerry [who was skinny] at the table: "I shall have a great advantage over you, Mr. Gerry, when we are all hung for what we are now doing. From the size and weight of my body I shall die in a few minutes, but from the lightness of your body you will dance in the air an hour or two before you are dead." This speech procured a transient smile, but it was soon succeeded by the solemnity with which the whole business was conducted....Benjn. Rush

22 Revolutionary Women and Children

George Washington said a wife should have "good sense, a good disposition, a good reputation, and financial means." Martha Washington had all of these.

Well, that Declaration did it! We Americans announced that we were free, and then we had to make it real. England wasn't going to give up her colonies without a fight. In 1775 King George had proclaimed that the colonies were in rebellion. But that Declaration of Independence in 1776 changed the nature of the conflict. It said that we no longer wanted to be colonists. This wasn't a little family squabble anymore. It had become a war to found a nation. It was war for a revolutionary idea: the idea that people could rule themselves. And so it was called the American Revolution. It was a people's war—and people means men, women, and children. It wasn't only the men who would do battle.

A British officer told his general that if all the men in America were killed, "We should have enough to do to conquer the women." One British soldier wrote home to England, "Even in their dresses the females seem to bid us defiance...on their shoes [they wear] something that resembles their flag of thirteen stripes."

Margaret Corbin was 23 when her husband went to war; she went with him. When he was killed, "Molly" Corbin took his cannon and kept firing.

Another Molly, Mary Hays, also helped fill her husband's place at a cannon. But she is most remembered for dodging shells as she carried a water pitcher to thirsty soldiers. She was known as Molly Pitcher.

Deborah Sampson was a soldier who disguised herself as a man. She served in the army for three years. Sampson was wounded twice, but took care of her own wounds to avoid being found out. Then she

Fighting Food

It was usually women (or children) who cooked for the soldiers. This is how to make biscuit, or hardtack—try it yourself. It doesn't have much to recommend it except a long shelf life, but all you need is flour and water.

Add enough water to some flour to make a soft (but not sticky) dough. Punch and work the dough for about 10 minutes (this will give you strong arm muscles). The dough will become like bubblegum: elastic. Roll it out 1/2" thick and cut into circles (use the floured rim of a glass). Prick with a fork and bake at 450° for 7 minutes. Turn the oven down to 350° and bake 7 to 10 minutes more. The biscuits should be hard as rock.

came down with a fever and ended up in a field hospital. That's where an amazed doctor learned the truth.

The doctor took Deborah Sampson to his house to care for her. When his niece decided she wanted to marry the handsome "soldier," the doctor decided he would have to tell Sampson's general the truth. In later years, Sampson went on a speaking tour telling of army life.

Anna Marie Lane was the only woman to receive a Revolutionary War soldier's pension from the Virginia Assembly. She enlisted in the army with her husband, but only he knew she was a woman. It wasn't that hard to keep it a secret. Soldiers rarely bathed, and they slept in their uniforms. Lane fought in four major battles until she, too, was discovered by an army doctor after being wounded.

How many women fought and were not discovered? Well, if they weren't discovered, and didn't write it down afterward, we'll never know.

Abigail Adams never stopped reminding her husband, John, about the inequality of opportunity for women in America compared with men. But he didn't listen.

"The men say we have no business with political matters," Eliza Wilkinson wrote to a friend, "[But] I won't have it thought that because we are the weaker sex (as to bodily strength my dear) we are capable of nothing more, than minding the dairy."

Although Molly Pitcher (left, in an artist's impression) may not really have helped fire a cannon, there were women who did, such as Margaret (Molly) Corbin.

Julia Stockton, the talented daughter of one signer of the Declaration of Independence, Richard Stockton, married another, Benjamin Rush. Their oldest son, Richard, went on to become U.S. minister in England and then secretary of the treasury under President James Monroe.

Most women stayed home during the war, but they did things they hadn't done before. They had to do all the men's work as well as their own. They ran farms and businesses, sewed clothes for soldiers, and helped make gunpowder and cannonballs. When battles were fought near their homes, women fed and cared for the wounded. Some women followed the army, acting as cooks and laundresses.

Children were part of it, too. An observer in Massachusetts watched "Children making Cartridges, running Bullets, making Wallets [soldiers' bags], and baking biscuit [soldiers' food]."

Women who were Loyalists had a terrible time of it. Usually they had to keep quiet or leave the country.

Molly Brant, Sir William Johnson's Mohawk widow, fought with the Iroquois on the side of Great Britain. It was said that "one word from her goes farther with [the Iroquois] than a thousand from any white man." Johnson had died in 1774. Some people believed that he, too, would have fought for the British. It was England that tried to honor the Indian treaties. Others said that if Johnson were alive it was he, not Washington, who would have been general of the armies.

Women and children acted as spies—for both sides. They were not likely to be suspected. Some women were innkeepers and listened as they served beer to the soldiers.

Many had to face the terrors of war on their own. Smallpox moved through the country as the armies moved. Soldiers carried the germs. Thousands died: soldiers, women, and children, too.

Women whose husbands had gone off to fight were often alone when the invading army looted and destroyed, as armies often do. Eliza Wilkinson's home in South Carolina was robbed by British soldiers. In her diary she wrote of a "day of terror," and of soldiers using "the most abusive language imaginable, while making as if to hew [cut] us to pieces with their swords." On Long Island, Lydia Mintern Post was forced to quarter enemy soldiers in her house. When they drank too much she said, "we have trying and grievous scenes to go through; fighting, brawls, drumming and fifing...and every abomination going on under our very roofs."

When it was necessary, women put guns to their shoulders. Those on the frontier were used to doing it. Many had fought Indians in those terrible raids in which right and wrong were often on both sides.

Foment means to cause something or to intensify its effect. ***Iniquitous*** (in-ICK-kwit-uss) means wicked or sinful. A ***scheme*** (SKEEM) is a plan or idea.

In Boston, Mercy Otis Warren wrote a play that mocked the British and championed the cause of revolution. She used her mind and wit to turn Loyalists into Patriots. So did Phyllis Wheatley. At age seven Wheatley had wept when she stepped off a slave ship; now she was writing patriotic poems that were praised by George Washington.

Martha Washington had never been outside Virginia when she got into her coach and headed for Massachusetts to join her husband. People cheered her along the way. She proved that she was made of strong fiber during cold months spent in army camps.

Eliza Pinckney saw much of her wealth disappear; the war was hard on the Pinckney plantations. Instead of complaining, she said she was probably better off with "moderation in prosperity." Some American women were independent already, although John Adams, Thomas Jefferson, and the other men in Philadelphia weren't ready to acknowledge it. That was a subject on which they were obtuse. (That word means thickheaded.)

When it came to equality for women, Adams and the other delegates ignored the subject.

John Adams had no excuse at all for being obtuse on women's equality. His wife, Abigail, kept telling him that all his talk about independence and freedom was a little strange if he couldn't understand that half of the population was not free.

And women weren't really free. They were ruled by their fathers or husbands. They couldn't vote. They had no representation.

Here is the way Abigail said it in a letter to John:

> *Whilst you are proclaiming peace and goodwill to men...you insist upon retaining an absolute power over wives. But you must remember, that arbitrary power is like most other things which are very hard, very liable to be broken.*

In another letter she wrote:

> *In the new code of laws...I desire you remember the ladies...if particular care and attention are not paid to the ladies we are determined to foment a rebellion and will not hold ourselves bound to obey any laws in which we have no voice or representation.*

Women weren't the only ones who weren't free. Abigail knew that. In another letter to John she wrote:

> *It has always appeared a most iniquitous scheme to me to fight ourselves for what we are daily robbing from those who have as good a right to freedom as we have.*

She was talking about slaves and slavery.

Phyllis Wheatley was freed by her owner, whose last name she bore.

Equal Rights

In 1774, a group of slaves from Massachusetts wrote a letter to Governor Gage demanding freedom. This is part of what they said:

We have in common with all other men a natural right to our freedoms...we are a freeborn people and have never forfeited this blessing by any compact or agreement whatever. But we were unjustly dragged by the cruel hand of power from our dearest friends and some of us stolen from...our tender parents and from a populous, pleasant and plentiful country and brought hither to be made slaves for life.

Massachusetts slaves went to court and won their freedom through a series of court decisions. In most other northern states, laws were passed ending slavery. Those laws were not always enforced.

23 Freedom Fighters

Everyone who knew Anthony Benezet described him as a man who did good and asked nothing in return. During the day he taught white children and at night he taught blacks. He soon discovered that the black and white children were equal in their abilities (many whites had a hard time believing that). Benezet did everything he could to get Philadelphia's slave owners to free their slaves. When he died, in 1784, hundreds of blacks mourned at his funeral.

Many black men took part in the battle of Breed's Hill. This one may be a soldier named Salem Poor, who fought very bravely.

James Forten had to beg his mother to let him go to sea. His father had died when he was seven and his mother depended on him. Besides, he was her pride and delight. James was bright and full of fun and willing to work hard and please. He'd gone to Anthony Benezet's well-known Quaker school until he was nine, so he could read and write well and handle numbers, too. Then he'd gone to work: first at Benezet's grocery store and then on Philadelphia's docks. His father had been a sailmaker (sewing the heavy sailcloth with waxed thread took special skill). James learned the trade and eventually became one of Philadelphia's most successful and wealthiest sailmakers.

But when James was 14 there was a war being fought, and he wanted to be a part of it. Like everyone in Philadelphia, he'd seen George Washington and Benjamin Franklin, and he'd heard those words, "all men are created equal." They were worth fighting for.

So he signed up on the *Royal Louis*, a privateer. The colonies had small navies, but they didn't amount to much against the great English fleet. It was the private ships that were hurting England. Congress allowed them to attack British ships and keep any profits that they made.

Powder boys were also called ***powder monkeys***.

James Forten became a powder boy. It was a dangerous job. Eighteenth-century ships were wooden, and flammable. Cannons were kept on deck, but gunpowder was stored below, where it was more likely to stay dry and be safe from accidents. When fighting began someone had to keep the cannons supplied. It was powder boys who did that. They needed to be small, fast, and fearless. They needed to keep running—up and down the ship's stairs, or companionways—even if cannonballs were falling around them and men were screaming in pain.

Which is exactly what happened on James Forten's first voyage out. The *Royal Louis* met a British ship; there was an awful fight with exploding shells, screams, groans, and deaths. But when the *Royal Louis* returned to Philadelphia, her guns were pointed at a captive British ship. It seemed as if all Philadelphia came out to cheer. James Forten knew what it was to be treated as a hero. And he got his share of the profits when the English ship was sold. Then he, and the other crew members who survived, repaired their ship and were soon back at sea. This time they were not lucky.

When the *Royal Louis* went after a British ship, two others appeared out of the blue. The Americans had stumbled into a trap; they were outnumbered; there was nothing to do but surrender. James Forten and his mates were brought aboard a British ship as prisoners. Now Forten had a special worry. In Philadelphia he was a free person, but he knew that the British often sold black prisoners to slave dealers in the West Indies—and James had dark skin. What would the ship's captain do with him? He was taking most of the American captives to a prison ship—and that was a bad enough fate. Would he take Forten with them? Or, even worse, would he sell him?

Later in his life Forten liked to tell the story of what happened next. He said it was his skill at playing marbles—learned as a boy in Philadelphia—that saved him from slavery. It so happened that the British captain had a son who was about Forten's age. When the English boy saw the young prisoner playing marbles, he asked to play. Forten beat him. But James Forten was so likable that the captain's son asked his father to take him back to England and set him free. The captain said he would do it, if James would renounce his country.

Forten wouldn't consider it. He was an American and he said, "No! I shall never prove a traitor to my country!" and that was that. He spent the next seven months in an overcrowded, stinking prison ship before

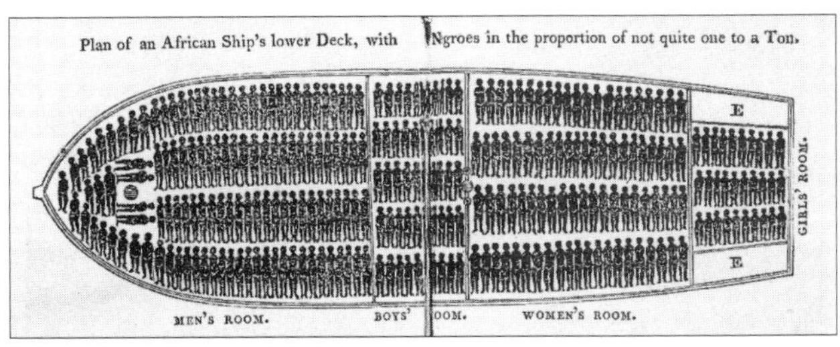

Slaves came to the colonies squeezed onto ships (such as the one shown above) in spaces about 15 inches wide. In America, they were expected to work hard for no pay.

Tom Paine asked how American slave owners could "complain so loudly of attempts to enslave them [by Britain], while they will hold so many hundred thousand in slavery."

This slave auction was unusual for being held indoors. Most were conducted right on the street.

What would you do if you were a black person living in Philadelphia at the time of the Revolution? James Forten wasn't typical. Many of Philadelphia's blacks supported the British. The English offered freedom to slaves (or so they said). But then there was that Declaration of Independence to think about. Black people understood the real meaning of those words "all men are created equal" long before most whites did.

he was released in a prisoner exchange.

No one knows the numbers for sure, but about 5,000 black men and boys are said to have fought on the American side during the Revolutionary War (out of about 300,000 soldiers in all). A redcoat wrote, "No regiment is to be seen in which there are not Negroes in abundance and among them are able-bodied, strong, and brave fellows."

Some helped England. The British promised to free any slaves who left their Patriot masters. (They didn't promise freedom to slaves of Loyalists. Those slaves were returned to their owners.) In the South it wasn't easy to escape to the British lines. There were lashings and other punishments for those who were caught. But tens of thousands of black men, women, and children "voted with their feet." They made it to British camps. Others were captured by the British. Thomas Jefferson wrote of what happened on his plantation:

"He [Lord Cornwallis] carried off about thirty slaves. Had this been to give them freedom, he would have done right; but it was to consign them to inevitable death from the smallpox and putrid fever, then raging in his camp. This I knew afterward to be the fate of twenty-seven of them. I never had news of the remaining three, but presume they shared the same fate." One of the slaves who died of fever was named Sam. He was nine years old. Another, Jane, was 10.

Many white Americans—especially in the South, where blacks often outnumbered whites—didn't want black people to have guns. They feared a slave uprising. So a year after the first battles a rule was made: no blacks—free or slave—in the Continental army. George Washington insisted on an exception to that rule. He said that those free black soldiers who had already fought should be allowed to reenlist. But otherwise blacks were excluded.

As the war went on—and on—some people began to see the nonsense of that policy. By the summer of 1777, New England's militias were using black soldiers again and promising them their freedom. Rhode Island had two black regiments. In Virginia, black soldiers fought alongside white soldiers. Most of the southern states used blacks as river pilots and even gunners on ships. Finally, in 1778, the Continental army changed its policy and began enrolling black soldiers. But South Carolina and Georgia still wouldn't go along.

Two young officers on General Washington's staff—John Laurens (from South Carolina) and Alexander Hamilton (from New York)—tried to get South Carolina to change its mind. They couldn't do it.

Why would blacks fight for a nation that allowed slavery? This war for freedom and equality was confusing. The words of the Declaration of Independence were noble—everyone agreed on that—but were they really meant to be taken seriously? Now that was where there was some disagreement.

The fact is, the whole idea of equality seemed insane to some people. Remember, the world of the 18th century was like a ladder, where everyone had a particular rung to stand on. It seemed secure, and sure. It was a world where everyone knew his or her place. You were always looking up, or down, at others. Equality would knock the ladder on its side. Equality is horizontal. Everyone is on the same level.

Many men and women weren't ready to make that change. (Children often have an easier time with that idea of equality than adults do.) Most grownups believed that if people didn't know their place, society would fall apart. And they were right. The society they knew—the society of the ladder—was going to be destroyed by that phrase "all men are created equal." No one could tell where it would lead, but some were scared.

This new idea—that people all have equal rights—seemed wild and radical. Equal rights? "Nonsense," said some. If equality were taken seriously it would change everything about people's relationships. It was going to happen, but not overnight. You needed to start with the idea. James Forten understood that. So did Thomas Jefferson. So did many others.

A Leading Citizen

After the war, James Forten became one of Philadelphia's most prosperous and influential citizens. He owned a large sailmaking business that employed both blacks and whites. He helped found an antislavery society (with Benjamin Rush) and was a leader in the African Methodist Church. He also supported women's rights and peace movements. When a white-owned firm closed (leaving large debts owed to Forten and others), James Forten went to the owner and said, "I came, sir, to express my regret at your misfortune....If your liabilities were all in my hands, you should never be under the necessity of closing your business." Forten acted as a bridge between the black and white communities, respected by all.

Philadelphia in 1765 had about 1,400 slaves and 100 free blacks. In 1783, Philadelphia had about 400 slaves and more than 1,000 free blacks.

24 Soldiers From Everywhere

Peace in Europe—well, that was almost true. Russia, Turkey, and Bavaria were waging minor wars. There was almost never total peace in 18th-century Europe.

These drilling soldiers are on a revolutionary recruiting poster. It took a foreigner, Baron von Steuben, to teach Americans to drill.

There was peace in Europe, and that was unusual. It was also a problem for European soldiers who knew no trade but fighting. So when the American Revolution began, many of Europe's soldiers knocked on Ben Franklin's door. Franklin was in France looking for help—financial help—for the American cause. He wasn't really looking for out-of-work soldiers, but he sent many of them to America anyway. When the European soldiers got to America, many of them wanted to be generals, or at least colonels. But American soldiers didn't want to fight under officers from other countries, especially those who couldn't even speak their language. That created some troubles.

So no one quite trusted the Marquis de Lafayette (mar-KEE-duh-laf-fy-ET) when he first landed in America. A marquis (you can say MAR-kwis or mar-KEE) is a French nobleman, like an English lord. This marquis was very rich, and noble in the best sense of the word. He was 19 years old.

His father had died fighting the British. Lafayette wished to avenge his father, and he also believed in the liberty the Americans were fighting for.

But in 1777, when he appeared in Philadelphia, he was taken for just another French soldier of fortune. "Thanks, but we don't need any more of your kind," was what he was politely told, in French, by James Lovell, who was chairman of the Continental Congress's committee of foreign applications.

Fighting together with men from other regions made many soldiers begin to think as Americans. General Nathanael Greene of Rhode Island spoke out against "local attachments." Greene said, "I feel the cause and not the place. I would as soon go to Virginia [to fight] as stay here [in New England]." This was a new way of thinking.

The marquis would not be dismissed. He had bought a ship to

come to America; he had paid for the soldiers who came with him; he had even angered his king, who was not yet ready to take sides in this war. Lafayette wrote a letter to John Hancock, president of the Congress. He asked for two favors: "The first is to serve at my own expense. The second is to begin my service as a volunteer."

Now that was an unusual request. John Hancock paid attention, and so did George Washington. And that brings us to one of the nicest stories of the war: the lifelong friendship of Washington and Lafayette. They became like father and son, and neither was ever to be disappointed in the other.

Like a young knight, Lafayette wanted to prove himself in battle, and he did. He became a general on Washington's staff, and suffered with the army during winter encampments. He fought well and, when he was wounded, his bravery endeared him to his men. He contributed much of his personal fortune to the American cause. Later, when he had a son, he named him George Washington.

General Andrew Lewis, describing soldiers in Williamsburg in 1776, said, "It is observed that many of the Soldiers when posted as Sentries take the liberty of sitting down. This unjustifyable practice is strictly forbid."

Lafayette was a noble man and a hero.

So was the Baron Friedrich von Steuben, even though he was a bit of an imposter. Von Steuben told Franklin that he had been a general on the staff of King Frederick the Great of Prussia.

That wasn't quite true. He was a captain in the Prussian army. Ben Franklin was not easily fooled; he saw through von Steuben right away. But he also saw that he was exactly what the army needed: a fine drillmaster. He thought von Steuben might be able to turn that untrained Continental army into a professional fighting force.

Franklin was right. Von Steuben had a happy personality, a lot of energy, professional knowledge of soldiering, and a roaring voice. He hollered at the American troops in a language that was a combination of German, English, and French with a few swear words thrown in. He made himself understood, and he did exactly what Franklin thought he would do. He turned a disorderly group of recruits into skilled soldiers. He trained them to fight as Europe's soldiers fought: with muskets and bayonets. He taught them to follow complicated orders and execute complicated maneuvers. He made them as good as the best British troops.

The Marquis de Lafayette became known as "the soldier's friend." Although he returned to France and is buried there, his grave is covered with earth from Bunker Hill.

He was also smart enough to discover that there was something different about American soldiers. They were independent men who wouldn't take orders unless they understood the reason for them. In Europe, von Steuben said, soldiers did what they were told. In

My Dear Heart

***When Lafayette said he wanted to go to America to fight for freedom, French officials wouldn't let him. So he made secret plans, put a black wig over his red hair, and set sail from Spain. He couldn't even tell his wife, Adrienne, goodbye. This is what he wrote her from his ship,* La Victoire.**

My dear Heart: It is from far away that I am writing, and added to this cruel distance is the still worse uncertainty as to when I shall have news of you....How will you have taken my going? Do you love me less? Have you forgiven me?... I shan't send you a diary of the voyage; days follow each other and are all alike; always sea and sky and the next day just the same....As a defender of Liberty which I adore...coming to offer my services to this interesting republic, I am bringing nothing but my genuine good will.

America, he said, soldiers wanted to know why an order was given; then they would do it.

Most of the soldiers who fought in the war were said to be Scotch-Irish. They, or their ancestors, had been poor farmers in Scotland who were lured to northern Ireland by English promises of cheap land. They'd been fooled. Life in northern Ireland was hard, so when they heard of the opportunities in America, a quarter of a million of them packed up, took their chances as indentured servants, and headed across the ocean. Because they had moved about, they quickly thought of themselves as Americans—not Scotsmen, or Irishmen, or Virginians, or Pennsylvanians. They were good soldiers.

Haym Salomon was not a soldier, but the help he gave the Revolution was as important as that given on any battlefield.

Salomon was Polish and a Jew. He had longed for religious freedom and liberty in his native country. But when Poland was invaded by Russia, Haym Salomon had to flee. When he came to America he felt at home.

Robert Morris had the impossible task of finding the money to pay for the war. He did it—but died penniless himself.

Salomon was a shy man who spoke several languages and had a talent for the language of business. He had a reputation for integrity; people trusted him.

The British shouldn't have trusted him. When they captured New York, Salomon spied on them for the Patriot cause. He was captured, imprisoned, paroled, captured again, and imprisoned again. Finally, he fled to Philadelphia. Soon, in his quiet way, he won the confidence of the French who had

Haym Solomons,

BROKER to the Office of Finance, to the Conſul General of France, and to the Treaſurer of the French Army, at his Office in Front-ſtreet, between Market and Arch-ſtreets, BUYS and SELLS on Commiſſion

BANK Stock, Bills of Exchange on France, Spain, Holland, and other parts of Europe, the Weſt Indies, and inland bills, at the uſual commiſſion.——He Buys and Sells

Loan-Office Certificates, Continental and State Money, of this or any other ſtate, Paymaſter and Quartermaſter Generals Notes; theſe and every other kind of paper tranſactions (bills of exchange excepted) he will charge his employers no more than ONE HALF PER CENT on his Commiſſion,

He procures Money on Loan

for a ſhort time, and gets Notes and Bills diſcounted.

Gentlemen and others, reſiding in this ſtate, or any of the united ſtates, by ſending their orders to this Office, may depend on having their buſineſs tranſacted with as much fidelity and expedition, as if they were themſelves preſent.

He receives Tobacco, Sugars, Tea, and every other ſort of Goods to Sell on Commiſſion; for which purpoſe he has provided proper Stores.

He flatters himſelf, his aſſiduity, punctuality, and extenſive connections in his buſineſs, as a Broker, is well eſtabliſhed in various parts of Europe, and in the united ſtates in particular.

All perſons who ſhall pleaſe to favour him with their buſineſs, may depend upon his utmoſt exertion for their intereſt, and——

Part of the Money advanced, if required.

N. B. Paymaſter-General's Notes taken as Caſh for Bills of Exchange.

F O R S A L E,

At the top of Haym Salomon's advertisement for his banking business are the words "Broker to the Office of Finance"—which meant that he lent money to the revolutionary government.

Ambassador Franklin pays his respects to King Louis of France. He and his co-minister, Silas Deane (inset), had to persuade France to help America.

come to the aid of America. The French made Salomon a general and their army paymaster.

Robert Morris trusted him, too. Morris, who was superintendent of finance for the Continental Congress, had the very difficult job of paying for the war. The colonies weren't much help. They raised very little money, and foreign countries didn't want to lend money to the Continental Congress. They doubted that the struggling new nation could beat mighty Britain, or pay its bills.

War and Peace Pipe

Normally, Thomas Jefferson didn't smoke. But he made an exception when he was with his Indian friends and they passed the pipe of peace. An Indian brother, a chief, wanted to know what this war was about. Jefferson, who was Virginia's governor, puffed on the pipe and then gave this explanation. What do you think of it?

Our forefathers were Englishmen, inhabitants of a little island beyond the great water, and, being distressed for land, they came and settled here. As long as we were young and weak, the English whom we had left behind, made us carry all our wealth to their country, to enrich them; and, not satisfied with this, they at length began to say we were their slaves, and should do whatever they ordered us. We were now grown up and felt ourselves strong; we knew we were free as they were, that we came here of our own accord and not at their biddance, and were determined to be free as long as we should exist. For this reason they made war on us.

General Washington with a group including three French officers: Lafayette (left), the Comte de Rochambeau (second from right), and, behind Rochambeau, the Marquis de Chastellux.

So Morris turned to Haym Salomon. Banks lent money to Salomon because they trusted him, even when they wouldn't lend it to the Congress. In Morris's diary you can count 75 times that he went to Salomon for help. Members of the Continental Congress needed help, too. Morris was unable to pay their salaries. James Madison, James Monroe, and Baron von Steuben were among those Salomon helped.

But mostly he helped the struggling army. Sometimes he dipped into his own pocket. When he had no more money to give he turned to the Jewish community and to others. The Jews were few in number, but the ideals of the new nation spoke to them in a special way because they had often been persecuted in the Old World. When he died in 1785, at age 45, Haym Salomon was almost penniless; he had given his country everything he owned.

Jack Jouett's Ride

Jack Jouett

After Thomas Jefferson wrote the Declaration of Independence, he went home to Virginia, and, a few years later, was elected governor. That was during a difficult time when the British army invaded Virginia. The redcoats wanted to capture Jefferson and the members of the Virginia General Assembly—and they almost did. (Note: a dragoon *is a British soldier.* Monticello *is the name of Jefferson's house.)*

Hardly anyone has heard of the ride
Of big Jack Jouett through the countryside;
No poet told of his frantic flight
Through Virginia's forest in the dark of night.

The British were marching, they were heading west
Seeking one prize over all the rest:
It was the man who had made the King glower,
Virginia's governor—WANTED—for London's Tower.

Big Jack, feather in his cap, cut by briars, short of sleep,
Had rivers to cross and fences to leap,
Till he reined in his horse and came to a stop
At a house, Monticello, on a mountain top.

"Dragoons," he warned. "They're coming, they're real!"
The governor, at breakfast, finished his meal.
Then he mounted his horse and rode off and away,
A minute later—that's the truth, so they say—
The redcoats arrived; too late, and they knew it,
Thomas Jefferson was gone, with thanks to Jack Jouett.

25 Fighting a War

This British cartoon poked fun at American soldiers, but the shabby, makeshift uniforms and hungry looks were close to the truth.

If ever you write a book of history, you'll find the hardest part is deciding what to put in the book and what to leave out. That's a problem I'm having right now. It has to do with the Revolutionary War. If I write about all the battles, and all the things the Continental Congress was doing, and all the important people—well, that would take several books.

So, what to leave out? To begin with, descriptions of some battles. It's not that they were boring, or unimportant. There just isn't room to tell about them all. (Of course, you can read about them on your own.)

To be honest, it may be that I am skipping some of the war because it was so dreary and cold and discouraging. After Bunker Hill and Sullivan's Island, nothing seemed to go right for the Americans. Just thinking about it is painful. Poor General Washington—no one else would have put up with all the hardships. In wintertime his soldiers almost froze—some actually did—and many didn't have shoes, or enough food to eat, or proper guns to use.

An act of Congress read: "the pay of officers and privates [is] as follows:

Privates...$6 2/3 per month
Sergeants...$8 " "
Captains...$20." "

Now $6 was worth a whole lot more then than it is today, but it still wasn't much. Out of that pay the soldiers were "to find their own arms and clothes." Congress couldn't even pay for guns or uniforms. And

Enemies?

Sometimes being on opposing sides in a war divided fathers from sons and brothers from brothers; sometimes it hardly seemed to matter at all. Washington wrote to his brother about one incident between soldiers:

One day, at a part of the creek where it was practicable, the British sentinel asked the American, who was nearly opposite him, if he could give him a chew of tobacco: the latter, having in his pocket a piece of a thick twisted roll, sent it across the creek to the British sentinel, who after taking off his bite, sent it back again.

Two of the German soldiers who fought for the British. These were jägers (YAY-ghers), former foresters who were often used as scouts.

The British troops marched from Long Island into New York City and occupied it for the rest of the war.

The Americans on Long Island were trapped by British troops drawn up between them and their retreat across the East River. They had to fight their way back through the enemy. One rebel said later, "When we began the attack, General Washington...cried out, 'Good God, what brave fellows I must this day lose!'"

those salaries hardly ever got paid. Congress seemed to spend its time talking and not doing much else. It just didn't have any money. Talk about frustration!

Most of the soldiers would have run off—deserted—if it hadn't been for their respect for General Washington. As it was, some did desert, and many others signed up for only three months. These were citizen-soldiers, not professionals like the Europeans. As soon as Washington got them trained, it was time for them to go home.

For all his cool under fire, Washington was said to have had a fierce temper. He must have had a hard time keeping it under control those first months after he took charge. Everything seemed to go wrong.

The British didn't just sit around and let the colonies rebel. They sent an army to put down the revolution. One day in the summer of 1776, a New Yorker named Daniel McCurtin looked out his window and saw the wooden masts of hundreds of British ships. He described what he saw:

> *I...spied as I peeped out...something resembling a wood of pine trees ...the whole bay was full of shipping...I thought all London was afloat.*

Those ships were full of soldiers; the British were landing an army in New York.

General Washington had his army in New York, too. But his men were inexperienced, and the British were not. On Long Island (a part of New York), the Americans marched into a trap: they were outnumbered, they panicked, they did many things wrong. The war might have been over right then, soon after it began, but Washington knew when to retreat and save his men. And he had some luck.

The luck came in the form of fog, thick fog. The cool general decided to move his troops from Long Island, across the East River to Manhattan Island. He had more luck: Massachusetts fishermen were manning his boats. Their eyes were used to fog. Before the British knew what had happened the Americans were across the river and saved to fight again.

Most of the soldiers the Americans fought were not even British. They were German—called "Hessians"—because many of them came from Hesse in Germany. Remember, Europe had been fighting wars for centuries. Large numbers of men in Europe spent their whole lives fighting. That was the only profession they knew, and they didn't much care who they were fighting for. Many didn't have a choice; they were forced to fight by their rulers. Some German princes made money by supplying soldiers to anyone who wanted to pay for them. The soldiers were called "mercenaries."

Thirty thousand German troops fought in America. Almost half never returned to Germany: some of them died, some chose to stay in the new country.

The Americans were furious that the British would hire soldiers to fight them. After all, most still thought of England as their mother country. Many Americans who were undecided about supporting the revolution became Patriots when they saw the mercenaries.

As the British scaled the cliffs on the New Jersey side of the Hudson River, "the rebels fled like scared rabbits," one Englishman said. They left "their artillery, stores, baggage and everything else behind them; their very pots boiling on the fire."

Total Surprise

General Howe, who commanded the redcoats on Long Island, was sarcastic about the Continentals' inexperienced tactics in preparing for battle:

Their [the Americans'] want of judgment had shone equally conspicuous during the whole of this affair. They had imagined... that we should land directly in front of their works, march up and attack them without further precaution in their strongest points. They had accordingly fortified those points with their utmost strength, and totally neglected the left flank...It was by marching round to this quarter that we had so totally surprised them on the 27th, so that the possibility of our taking that route seems never to have entered their imaginations.

26 *Ho*we Billy Wished France Wouldn't Join In

General Howe had already served in America. In 1759 he led Wolfe's troops to seize Quebec.

Sir William Howe (who was sometimes called Billy Howe) was in charge of all the British forces in America. It was Howe who drove the American army from Long Island to Manhattan. Then he chased it across another river to New Jersey. And, after that, he forced George Washington to flee on—to Pennsylvania. It looked as if it was all over for the rebels. In New Jersey, some 3,000 Americans took an oath of allegiance to the king. But Washington got lucky again. The Europeans didn't like to fight in cold weather.

A ***hoop-stay*** was part of the stiffening in a skirt; a ***jupon*** was part of a corset. ***Matrons*** are married women. The ***misses*** are single girls, and ***swains*** and ***beaux*** are young men or boyfriends. ***Making love*** meant flirting.

Sir William settled in New York City for the winter season. Howe thought Washington and his army were done for and could be finished

Swarming with Beaux

Rebecca Franks was the daughter of a wealthy Philadelphia merchant. Her father was the king's agent in Pennsylvania, and the family were Loyalists. Rebecca visited New York when it was occupied by the British. Her main interest in the war was that it meant New York was full of handsome officers:

My Dear Abby, By the by, few New York ladies know how to entertain company in their own houses unless they introduce the card tables....I don't know a woman or girl that can chat above half an hour, and that on the form of a cap, the colour of a ribbon or the set of a hoop-stay or jupon....Here, you enter a room with a formal set curtsey and after the how do's, 'tis a fine, or a bad day, and those trifling nothings are finish'd, all's a dead calm till the cards are introduced, when you see pleasure dancing in the eyes of all the matrons....The misses, if they have a favorite swain, frequently decline playing for the pleasure of making love.... Yesterday the Grenadiers had a race at the Flatlands, and in the afternoon this house swarm'd with beaux and some very smart ones. How the girls wou'd have envy'd me cou'd they have peep'd and seen how I was surrounded.

off in springtime. Besides, Billy Howe loved partying. And some people say he liked the Americans and didn't approve of George III's politics. For reasons that no one is quite sure of, General Howe just took it easy.

But George Washington was no quitter. On Christmas Eve of 1776, in bitter cold, Washington got the Massachusetts fishermen to ferry his men across the Delaware River from Pennsylvania back to New Jersey. The river was clogged with huge chunks of ice. You had to be crazy, or coolly courageous, to go out into that dangerous water. While the Massachusetts boatmen were getting the army across, the Hessians, on the other side—at Trenton, New Jersey—were celebrating Christmas by getting drunk. Before the Germans could focus their eyes, the Americans captured 900 of them.

A week later, Washington left a few men to tend his campfires and fool the enemy. Then he quietly marched his army to Princeton, New Jersey, where he surprised and beat a British force. People in New Jersey forgot the oaths they had sworn to the king. They were Patriots again.

Those weren't big victories that Washington had won, but they certainly helped American morale. And American morale needed help. It still didn't seem as if the colonies had a chance. After all, Great Britain had the most feared army in the world. It was amazing that a group of small colonies would even attempt to fight the powerful British empire. When a large English army (9,500 men and 138 cannons) headed south from Canada in June 1777, many observers thought the rebellion would soon be over.

The army was led by one of Britain's

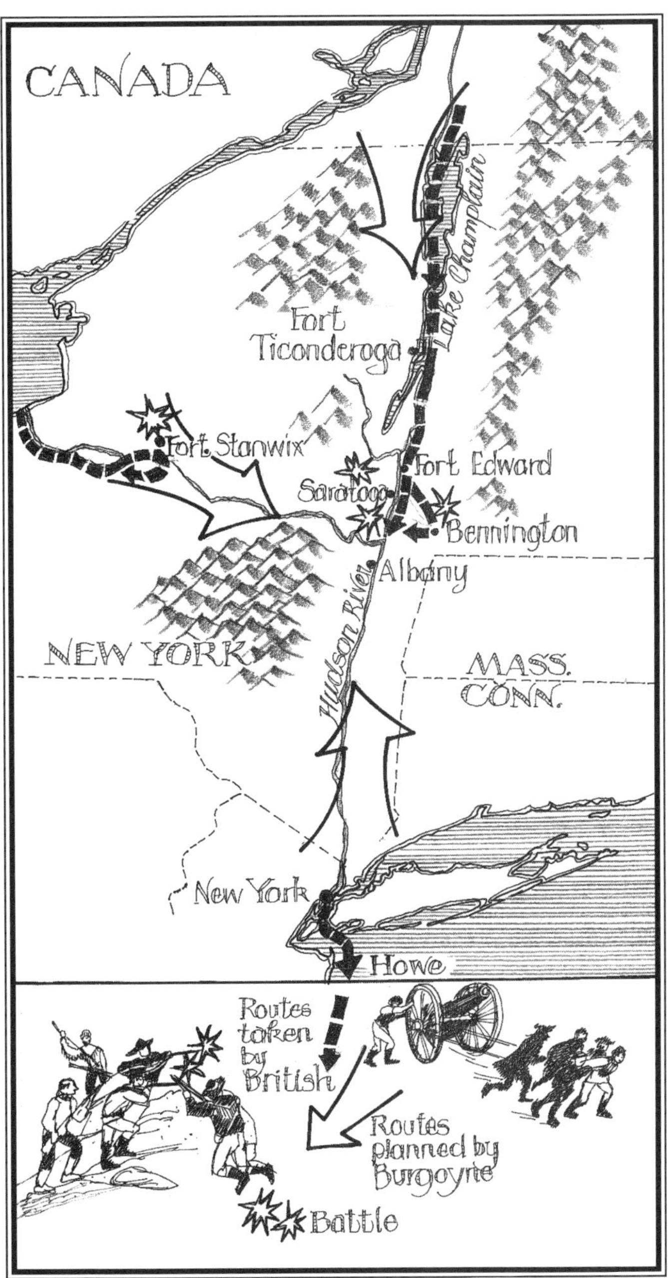

General Burgoyne's redcoats carried far too much equipment. Each man's boots alone weighed 12 pounds. They took two months to cover 40 miles from Fort Ticonderoga to Saratoga, and lost hundreds of men to American snipers.

General Gates

Four Plates and Two Glasses

A Boston newspaper tells of the British surrender at Saratoga:

General Gates invited General Burgoyne and the other principal officers to dine with him. The table was only two planks laid across two empty beef barrels. There were only four plates for the whole company. There was no cloth, and the dinner consisted of a ham, a goose, some beef and some boiled mutton. The liquor was New England rum, mixed with water, without sugar; and only two glasses, which were for the two Commanders-in-Chief; the rest of the company drank out of basins....After dinner, General Gates called upon General Burgoyne for his toast which embarrassed General Burgoyne a good deal; at length, he gave General Washington, General Gates, in return, gave the King.

General Burgoyne

most colorful officers, General John Burgoyne. Burgoyne was known as Gentleman Johnny. He was a wild character: a drinker, gambler, actor, playwright—and a pretty good general. He said:

> *I have always thought Hudson's River the most proper part of the whole continent for opening vigorous operations. Because the river, so beneficial for conveying all the bulky necessaries of an army, is precisely the route that an army ought to take for the great purpose of cutting the communications between the Southern and Northern provinces.*

Which means, in ordinary English, if Burgoyne could capture the Hudson River area, he would cut off New England and New York from the rest of the colonies.

Burgoyne made careful plans. His army would go south along Lake Champlain and the Hudson River, heading for Albany, New York. Sir William Howe was supposed to come north, along the Hudson, from New York City to Albany. Another British army was expected from the west. Burgoyne planned to trap the American army like an insect squashed between three fingers. Then he and Howe could march south together and mop up the rest of the rebel forces.

Things didn't go as Burgoyne expected. General Howe decided to head for Philadelphia instead of Albany. And the western army got involved in other battles and never made it east. Nevertheless, General Burgoyne went ahead with his plans. He sailed down Lake Champlain and recaptured Fort Ticonderoga (the fort that Ethan Allen, Benedict Arnold, and the Green Mountain Boys had taken). In London there were wild celebrations when that news arrived. King George yelled, "I have beat them! I have beat them!"

Then Gentleman Johnny went on to Fort Edward. When the Patriots saw him coming, they abandoned the fort. But they found another way to fight. They cut down trees and threw them all over the roads. That slowed the British army and the wagons and the heavy cannons they dragged with them. The Americans sniped—Indian-style—from the woods. Those small guerrilla attacks kept the British soldiers on edge, and scared.

Burgoyne began running low on food. When he learned there were food and horses in Bennington, Vermont, he sent some troops to get them. That was a mistake. His men got whipped in a battle at Bennington. But, finally, the British army reached Saratoga, New York, and there Burgoyne faced a big decision.

Saratoga edges the Hudson River, just north of Albany. General

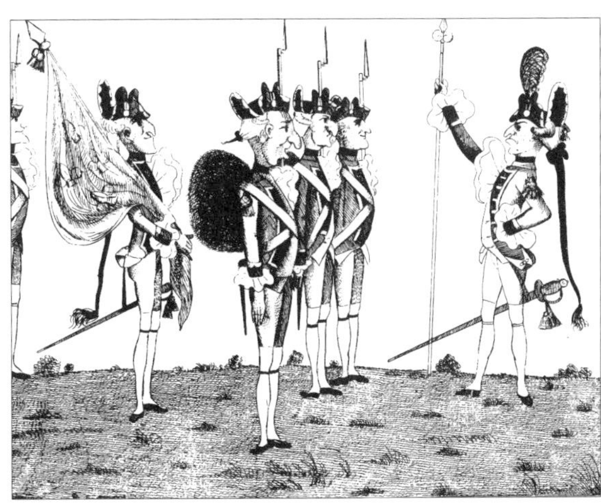

A British cartoon made fun of the fancy uniforms and hairdos of the French soldiers.

Horatio Gates was in command of the American forces at Saratoga. Gates placed his men on a high bluff overlooking the road to Albany where it squeezes between hills and river. Colonel Thaddeus Kosciuszko (kush-CHOO-shko), a Polish military engineer serving with the Americans, had picked the site and fortified it with cannons. That gave Burgoyne a difficult choice. He could march his men down that treacherous narrow road, or he could fight the Americans on their fortified heights. He chose to fight.

What happened? Farmers poured into the area; soon the American force was three times the size of Britain's army. The American farmers were sharpshooters, and their rifles were deadly accurate. The battle wasn't even close. The British lost about 600 soldiers at Saratoga; American casualties were about 150.

On October 17, 1777, the incredible occurred: the great General Burgoyne and his whole army surrendered! The European soldiers were marched to Boston, made to promise they would not fight in America again, and sent back to England. The Americans were jubilant.

A fast ship sailed out of New York harbor heading for France with word of the victory at Saratoga. This was just the kind of news Ben Franklin had been waiting to hear.

Many jokes and jeers were made after Saratoga at the expense of the British:

Burgoyne, alas, unknowing future fates,
Could force his way through woods but not through Gates.

Franklin had been sent to France to ask for aid. The French admired Franklin and they had a grudge against the British. They'd been fighting England off and on for hundreds of years. They were still mad about losing the French and Indian War. But they didn't think the colonists had a chance against Great Britain. No one wants to support a losing cause.

The victory at Saratoga changed everything. It got the French to join the war on the American side. France sent gunpowder—lots of it—and officers and soldiers and ships. (The next time you see a Frenchman or a Frenchwoman, you can say thank you. We might not have won independence without French aid.)

But the war wasn't over. There was still much hard fighting to be done.

At Christmas, Washington crossed the ice-choked Delaware River and captured nearly 1,000 Hessian troops in Trenton, New Jersey. He lost only two men, who froze to death.

The Love of His Men

From Baron von Steuben's instructions to his company officers:

A captain...must pay the greatest attention to the health of his men, their discipline, arms, accoutrements, ammunition, clothes and necessaries. His first object should be to gain the love of his men by treating them with every possible kindness and humanity, inquiring into their complaints and when well founded, seeing them redressed. He should know every man of his company by name and character. He should often visit those who are sick, speak tenderly to them, see that the public provision, whether of medicine or diet, is duly administered, and procure them besides such comforts and medicines as are in his power.

Baron von Steuben

27 Valley Forge to Vincennes

In bad weather, even muskets spent the night under cover, in little tents like these.

Things weren't going well for George Washington. He lost two battles in Pennsylvania: one at Brandywine, the other at Germantown. Then Sir Billy Howe captured Philadelphia, and that meant that Congress—which was meeting in the State House (now called Independence Hall)—had to flee. The fall of 1777 turned to winter. Howe was warm and comfortable in Philadelphia, as he had been the winter before in New York. He was partying again. Loyalist families entertained him and his men. Eighteen miles away, Washington, Lafayette, Baron von Steuben, and the American soldiers were miserable.

Washington had brought his army to a place called Valley Forge. It had been named for a nearby iron foundry, although the foundry was now in ruins—the British had destroyed it. Valley Forge was a good site from a military point of view. The land was high, near enough to Philadelphia to keep watch on that city, but not so close that the British could cause trouble with surprise raids.

There was little there, except for farmland and the Schuylkill (SKY-kull) River. There were no buildings for the army to use as barracks, and, in December when they arrived, the ground was covered with snow. The men had marched a long distance, and many were in rags. Within a few days the river turned ice hard. A cold wind began blowing. The soldiers pitched tents and started building huts of sticks, logs, and mud plaster. Washington, who was precise and cared about appearances, insisted that they all be the same size.

Picture 2,000 dirt-floored, drafty wooden huts lined up in streets like a village, and you have an idea of the architecture at Valley Forge. If you look at the ground, you may see blood. Some of the soldiers had no shoes, and their toes froze and left bloody tracks. Now add hunger to the scene, and you begin to get an idea of that terrible winter. But that was not the worst of it. Disease swept the camp. About 2,000 soldiers died.

That is the way it was at Valley Forge. There wasn't enough clothing. There wasn't enough to eat. It was fiercely cold. The officers feared a mutiny, and there were desertions. But not many. Most of those who slipped over to the British in Philadelphia were newcomers to the colonies.

There were no battles fought at Valley Forge. None at all. But something astounding happened there. A spirit evolved. It was amazing; the men who made it through that winter were better for it. They became a team: strong, confident, and proud of themselves, their country, and their leaders.

Lord—Lord—Lord

Dr. Albigence Waldo of Connecticut, a surgeon serving at Valley Forge, wrote about the Continental army's misery in his diary:

Dec 12th We are ordered to march over the river—it snows—I'm sick—eat nothing—no whiskey—no baggage—Lord—Lord—Lord. The army were till sunrise crossing the river—some at the wagon bridge and some at the raft bridge below. Cold and uncomfortable....

Dec 14th Poor food—hard lodging—cold weather—fatigue—nasty clothes—nasty cookery—vomit half my time—smoked out of my senses—the Devil's in it—I can't endure it—why are we sent here to starve and freeze....Here comes a bowl of beef soup—full of burnt leaves and dirt, sickish enough to make a Hector spew—away with it Boys—I'll live like the chameleon on air.

Washington tries to cheer his troops at Valley Forge. Benjamin Rush was shocked at the conditions he saw: "The troops dirty, undisciplined, and ragged...bad bread; no order; universal disgust."

Liberty Lover

It was lucky for the rebellious American colonies that Thaddeus Kosciuszko had a broken heart. You see, when he tried to elope with the girl he loved, her father wouldn't allow it. Kosciuszko hoped to forget the girl by coming to America to fight for freedom. He arrived just in time to join the American army in the battles of Fort Ticonderoga and Saratoga; then he headed south to do battle in the Carolinas.

Thaddeus Kosciuszko

Kosciuszko was more than a warrior. He was a fine thinker who cared about liberty. He became a friend of Thomas Jefferson. After the Revolutionary War, the grateful nation gave him United States citizenship and 500 acres of land in Ohio. But he wasn't ready to settle down. He returned to Europe to fight for freedom in Poland and France and Russia. Kosciuszko also freed all the serfs (who were almost like slaves) on his Polish estate (and that left him poor and in debt). In his will he asked that his American land be sold and the money used to buy freedom for slaves.

George Washington had a lot to do with that. At first Washington lived in a tent, among his men, and put up with hardships as they did. Later, he made his headquarters in a nearby four-room stone house, but the soldiers remained awed by his example. A young Frenchman who was there wrote of General Washington,

> *I could not keep my eyes from that imposing countenance....Its predominant expression was calm dignity, through which you could trace the strong feelings of the patriot, and discern the father, as well as the commander of his soldiers.*

Martha Washington could often be seen with a basket in her arms, bringing food and socks and cheer to those who needed it. Von Steuben made a difference that cold winter, too. He began training 100 men at a time and soon had the whole army drilled. The Americans were astonished. British officers did not conduct drills; they left that to their sergeants. This man did the drilling himself. He seemed to thrive on hard work, and nothing upset him. He was always good-humored, even when he was shouting and swearing.

Before long the Americans could march and maneuver, load and fire, use bayonets, and respond to complicated orders. But that wasn't enough for von Steuben. He expected them to be neat and shaved. Even rags, he told them, could be clean. He made all the officers set their watches by the same clock. He was determined that this army was going to be precise and proud of itself. Soon it was just that.

Washington appointed Nathanael Greene as quartermaster general. The quartermaster is in charge of supplies. Greene protested that he didn't want the job, but General Washington knew what he was doing. Greene brought enormous energy and determination to everything he did. He tramped around the countryside, found big caches of food and supplies, and hauled them to Valley Forge. By spring there was plenty of food, and clothing too.

"God grant we may never be brought to such a wretched condition again!" wrote quartermaster Nathanael Greene of Valley Forge.

In June the British left Philadelphia and headed for New York. They'd had a pleasant winter, but they hadn't accomplished a thing. The men who had gotten through the winter at Valley Forge were now a strong fighting force. They knew they could endure almost anything. They were ready to follow George Washington wherever he led.

While all that was going on at Valley

Forge, the Indians, who were being paid by the British for American scalps, were creating havoc on the frontier. (The Americans paid for British scalps.) But no matter which side they chose, the Native Americans would be losers. Their land was being taken from them. The European way of life and the Native American way of life seemed incompatible.

A ***cache*** is a hiding place for provisions, arms, or any kind of treasure.

Most of the settlers didn't understand what was happening. When they heard of Indian raids and scalpings, they were horrified. They believed what they had been told—that Indians were savages. They knew the English were signing treaties that gave Native Americans protection and rights; that was another reason they wanted the British out.

Things that are ***incompatible*** can't get along together.

George Rogers Clark took Kaskaskia with 175 men. It surrendered in 15 minutes.

So most colonists thought of Indian fighters as great heroes. Those Indian fighters, like George Rogers Clark, believed they were doing the right thing. Mostly they just wanted to push the Native Americans west, to free new lands for the settlers.

Clark was a frontiersman and a Patriot, as well as an Indian fighter. Born in Virginia, near Jefferson's home, Clark knew Indians well and could talk to them in a way they understood: he had learned to use their form of oratory. They called him Mitchi Malsa, which means Big Knife.

Clark was smart; he was also brave and daring. He had hardly any schooling, but he read all the books he could find. Some people called him the "Washington of the West." Like George Washington he was tall, very strong, and a surveyor—but that was where the resemblance ended. He had none of the dignity of the Virginia planter. Clark's personality swung from fierce temper to calm persuasion, but rarely rested anywhere.

He was 25 years old in 1778, when he persuaded Virginia's governor, Patrick Henry, to let him gather a force to take the Ohio Valley from the British and their Indian allies. Then he proceeded to win some astounding battles.

Here is some of what he did: with just 175 men and a few barges, Clark captured three strategically located British forts: Cahokia, Kaskaskia, and Vincennes. Then he talked the French inhabitants of the region into coming over to the American cause.

It was the battle of Vincennes that made him famous. When Clark fought at Fort Vincennes, he had only 150 men, most of whom were sick with chills and fever.

"A desperate situation," he said, "needs a desperate resolution." Clark sent a note to the British commander demanding the surrender of the fort. The British leader refused.

Clark attacked. He decided to confuse his enemy. He kept his men yelling like maniacs and demons as they fired through gun holes into the fenced fort. To those inside, it seemed as if a huge army was attacking. The fort depended on cannons for protection, but cannons are useless against moving targets. Clark never stopped moving. The British surrendered.

Later, England moved back into much of that region. But, for a while, it seemed as if a young backwoodsman had captured the Ohio territory. The Americans were frustrating the British. Men like George Rogers Clark just wouldn't fight the way they were expected to fight—and they never seemed to give up.

The Battle of the Kegs

What were the Americans to do about the British ships in the river near Philadelphia? David Bushnell was sure to figure out something, thought Colonel Joseph Burden, who was in charge of the American forces in Philadephia. Bushnell had already invented a submarine; maybe he could do something about those British ships. The colonel was right. Bushnell came up with a simple plan. He filled wooden barrels (called kegs) with gunpowder. Then he had the barrels put in the river. He expected the kegs to bump into the British ships and wham!—that would trigger an explosion.

Nothing happened. It looked as if the kegs would float harmlessly past the British ships. And they would have, if if hadn't been for some curious British sailors. They hauled a few of the kegs onto a barge—where they did explode. Actually, they didn't do much harm, but they certainly caused a commotion. "The alarm and consternation of the British was extremely great—the Military of every kind & order was seen in an instant, running in every degree of confusion and in every direction," said a report of the day. The Americans found it all funny, especially after Francis Hopkinson (a signer of the Declaration of Independence) told the tale in a tongue-in-cheek poem. With mock seriousness, he described the heroic British fight against some wooden kegs. Here is the last stanza of his poem:

Such feats did they perform that day
Against those wicked kegs, sir,
That years to come, if they get home,
They'll make their boasts and brags, sir.

The British, it seems, weren't poetry lovers. Soon afterwards they burned Hopkinson's house.

David Bushnell's submarine Turtle. *It failed to sink the British ship* Eagle *in 1776—but the British were so surprised that they fled anyway.*

28 The States Write Constitutions

George Mason was a bit of a loner; he had a low opinion of people's ability to get anything done by committee.

Most Americans wanted to be free of British rule. On that they now agreed. But exactly what kind of rule would they have after the British left? Who was going to be in charge? With a war going on, few people had time to think about what would come next. It was liberty that everyone was talking about.

But the men at the Continental Congress knew that someone had to plan for the future, so they suggested that each state write a constitution. Most of the states had been royal colonies with royal governors and royal charters. Now they needed new rules and new governors.

The states produced some fine constitutions—and a few that weren't so fine. And they learned things that were helpful later, when they wrote a constitution for the new nation.

State leaders spent a lot of time worrying about power. They didn't want anything in America like a too-powerful English king or a too-powerful Parliament. So they drafted state constitutions that divided power between a state congress (sometimes called an assembly), a governor, and law courts. They called it separation of powers. Think of a tree with three main branches: the assembly is the *legislative* branch, the governor is the *executive* branch, the courts are the *judicial* branch. The constitution writers tried to balance power so that no branch would have more weight than the others.

In 1780 Massachusetts found a good way to write its constitution. It elected special people to do it at a constitutional convention. Then all

Here reason shall new laws devise And order from confusion rise.

—Philip Freneau, American poet (1752–1832)

New Hampshire drafted its first constitution in January 1776, but a final version that pleased most of its citizens wasn't approved until 1783, after several tries.

the citizens of the state voted on the constitution. That may not sound unusual, but Massachusetts was the first place in the world to write a constitution that way. The Massachusetts Constitution began, "All men are born free and equal." Slaves began appearing before the Massachusetts courts, asking if those words meant what they said: all men are born free and equal. The courts always said yes. John Adams watched one such case. "I have heard there have been many," he wrote in his diary.

Note those words, "all men." It was more than a century before "all women" were given much thought, except in New Jersey, where the new constitution gave anyone who owned property the right to vote. Some said it was a printer's error, but women and black people voted. Later New Jersey took back that right. (A cheer, then a boo, for N.J.)

When Americans wrote their state constitutions, they argued about

- *freedom of speech and the press*
- *the right of the majority to change the government*
- *freedom of religion (which they called freedom of conscience)*
- *free education*
- *voting rights*
- *slavery*

That arguing didn't bring freedom to all. But the discussions were the beginning of a new way of thinking about government, and they were extraordinary for the times. Still, most states gave the vote only to white men who owned land. In some states Roman Catholics, Jews, Baptists, and atheists were barred from voting or holding public office.

It was a time of experiment.

Each state constitution had a Bill of Rights. Virginia's bill, written by George Mason, was a model for many others. It said that all government power was "derived from the people." Elections were to be free and citizens were not to be taxed

The first draft of Georgia's constitution was written by one of the signers of the Declaration of Independence, Button Gwinnett. Gwinnett was killed in a duel shortly afterward. Because Gwinnett died so soon after signing the Declaration, his signature is very rare. Autograph collectors have paid hundreds of thousands of dollars for an example.

Button Gwinnett

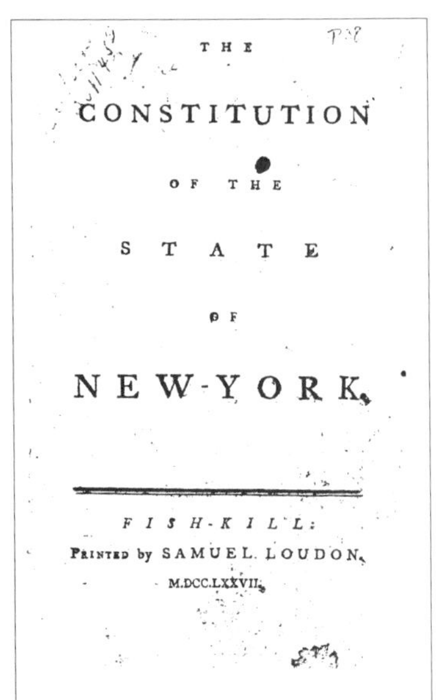

THE CONSTITUTION OF THE STATE OF NEW-YORK.

FISH-KILL: PRINTED by SAMUEL LOUDON, M.DCC.LXXVII.

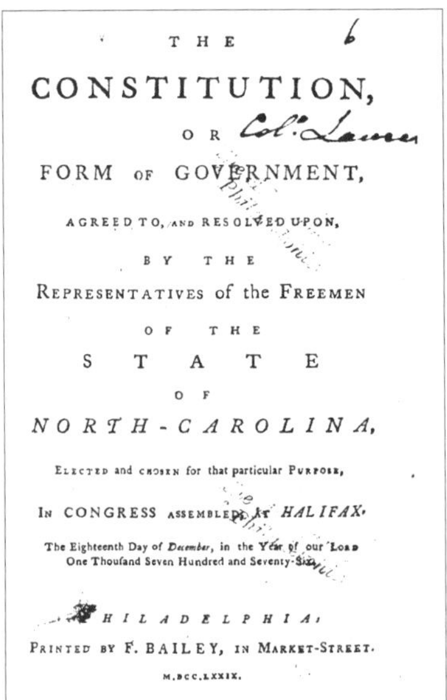

THE CONSTITUTION, OR FORM OF GOVERNMENT, AGREED TO, AND RESOLVED UPON, BY THE REPRESENTATIVES of the FREEMEN OF THE STATE OF NORTH-CAROLINA, ELECTED and CHOSEN for that particular PURPOSE, IN CONGRESS ASSEMBLED, AT HALIFAX. The Eighteenth Day of December, in the Year of our LORD One Thousand Seven Hundred and Seventy-Six.

PHILADELPHIA: PRINTED BY F. BAILEY, IN MARKET-STREET. M.DCC.LXXIX.

Lawyer John Jay drafted New York's constitution; he included the entire text of the Declaration of Independence. This copy of North Carolina's constitution was given to Benjamin Franklin.

(7)

refrained from oppreſſion, the people have a right, at ſuch periods as they may think proper, to reduce their public officers to a private ſtation, and ſupply the vacancies by certain and regular elections.

VII. THAT all elections ought to be free; and that all free men having a ſufficient evident common intereſt with, and attachment to the community, hath a right to elect officers, or be elected into office.

VIII. THAT every member of ſociety hath a right to be protected in the enjoyment of life, liberty and property, and therefore is bound to contribute his proportion towards the expence of that protection, and yield his perſonal ſervice, when neceſſary, or an equivalent thereto: But no part of a man's property can be juſtly taken from him, or applied to public uſes, without his own conſent, or that of his legal repreſentatives: Nor can any man who is conſcientiouſly ſcrupulous of bearing arms, be juſtly compelled thereto, if he will pay ſuch equivalent: Nor are the people bound by any laws, but ſuch as they have in like manner aſſented to, for their common good.

IX. That in all proſecutions for criminal offences, a man hath a right to be heard by himſelf and his Council, to demand the cauſe and nature of his

This was John Dickinson's copy of the Pennsylvania constitution. He didn't agree with much of what it said, and covered his copy with the changes he wanted made.

"without their own consent, or that of their representatives." Then it guaranteed every free person's right to a jury trial and to protection against unreasonable arrest.

Those were among the English rights that most Americans were determined to have in the new nation. The Virginia bill added something new. It was religious freedom. George Mason wrote a guarantee of religious tolerance. A young friend of Thomas Jefferson, named James Madison, suggested that the word "tolerance" be changed to "the free exercise of religion." Do you see why those words made a big difference?

Today we wish they had thought more about rights for every person—male and female, black and white, Indian and Asian. But, remember, the idea that government should guarantee freedom and equal opportunity in written documents was totally new. No nation had even tried to do it before. There were no guidelines to follow. This was a first step. Much of America's history from this point on would be an experiment. We would try new ideas—and make some terrible mistakes—but the goal would remain: freedom and equality for all.

If all this seems boring, it really isn't. Americans were doing something very exciting—and they knew it. Ordinary citizens were planning their own kind of government, putting it down in written documents, and then voting on it. England didn't even have a written constitution. People were always talking about the British Constitution, but what they meant was the collection of laws and court cases that had developed over centuries. The American state constitutions were unique. Here is what James Madison wrote of them: "It is the first instance, from the creation of the world...that free inhabitants have been seen deliberating on a form of government."

What does unique *mean?* Deliberating?

29 More About Choices

It was Henry Knox who got all the American guns across the icy Delaware. "Perseverance," he said, "accomplished what at first seemed impossible."

Decisions...decisions...they aren't easy. History has so many interesting stories, especially where we are, right now, near the end of the 18th century.

There just isn't room in this book to tell you about the first submarine built in America, or that it was tested during the American Revolutionary War. If you want to know about it, you'll have to find the story on your own. If you've been reading carefully, you'll know the name of the inventor.

I will tell you that some interesting things were happening in California about this time. The Russians had headed down the West Coast of the American continent and had built a small fort just north of what would someday be called San Francisco Bay. It looked as if the Russians might settle in California. The Spaniards didn't want that to happen. So they sent priests and settlers and built missions and ranches and farms. They began to create a new California. It was a gracious and satisfying place for the Spanish-speakers who were in charge, but mostly terrible for the Native Americans, who watched as their old world was destroyed.

At the Franciscan mission at Carmel, California, in 1786, Indians, friars, and the Spanish explorer Malaspina welcome a French explorer, La Pérouse.

Of course, no one in California, or in other Spanish-speaking settlements in New Mexico, Arizona, and Texas, would have believed you if you had told them that they would someday be part of a great nation that was beginning on the East Coast. That was hard to imagine. No one (except

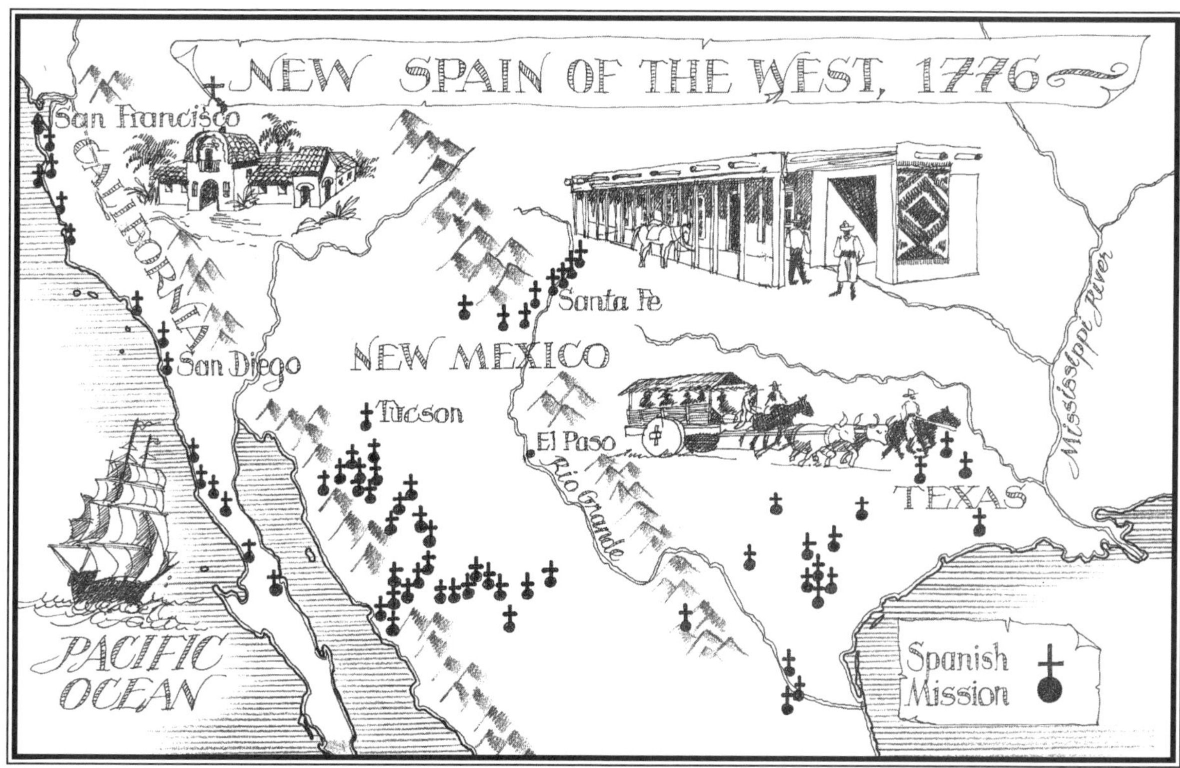

the Native Americans) had any idea of what most of the land was like between the coasts.

The Spanish-speakers heard tales of that battle for power in the English colonies, but it didn't seem to have anything to do with their world. They were wrong: the idea of independence was contagious. Soon the Spanish-speakers in Mexico wouldn't put up with a foreign nation dictating to them either. In the 19th century they broke away from Spain.

But right now we need to get back to that struggle on the East Coast. I want to tell you the stories of two people some historians forget. One story is about a man who married a woman from a Loyalist family. The war tore the family apart. That happened often, and it wasn't easy on anyone. The other story is about a fine printer, who happened to be a woman. Because of that, she lost a job, which wasn't fair at all.

Henry Knox was 12 when his father died; it was up to Henry to support his mother. He got a job in a Boston bookstore. Ten years later, he owned his own bookstore.

That was a good business for Henry Knox, because he loved to read. He especially liked reading about guns and cannons and other forms of artillery. He was fascinated with military life, so he enlisted in a militia,

While the English colonists were demanding religious freedom on the East Coast, the Spanish were trying to spread Catholicism in the West.

Lucy Flucker Knox spent the war years handling her own affairs. In 1780, she wrote to her husband that she expected to continue doing that in peacetime. "I hope you will not consider yourself as commander-in-chief of your own house," wrote Lucy Knox, "but be convinced... that there is such a thing as equal command."

It was an English poet named John Milton who made many Americans care about freedom of the press. Milton wrote a letter about a free press (back in the time of Oliver Cromwell and the English Civil War), and that letter was so powerful that it made people realize that in order to be free to think for yourself you need information. Information is what a free press gives you.

Knoxville, Tennessee, is named for Henry Knox. So is Fort Knox, Kentucky.

the Boston Grenadier Corps. A militia is a volunteer force of citizen soldiers. When Henry wasn't working at the bookstore, he trained and studied military affairs with the Grenadier Corps.

Some people said it was his militia uniform that made Lucy Flucker fall in love with him. But maybe it was his energy and intelligence that attracted her. Knox fell in love with her, too. Lucy's parents didn't approve, so they married secretly. Lucy's father, Thomas Flucker, was an official of the British government. He didn't think much of young Henry's politics. Henry wanted to see the colonies free of British rule.

When the Revolution began, Henry Knox volunteered. The Patriots were glad to have him. He knew a lot about artillery—even though it was all book learning. Henry was made a colonel and put in charge of artillery. It wasn't a big job; the Americans had very little artillery. (Artillery is cannons—big guns.)

Then Ethan Allen captured the British cannons at Fort Ticonderoga in New York. (You already know about that.) But the cannons weren't needed in upstate New York. The Continental army needed them in Boston, where they could fire at British ships in the harbor. No one was quite sure how to move them, or even if they could be moved. The cannons were made of iron or brass, and were very, very heavy.

Henry Knox was a happy-natured fellow who rarely got discouraged. Some people called him Knox the Ox, because he was huge. Well, Knox the Ox got real oxen—great strong animals—and put the cannons on sleighs, and started pulling them across rivers and over mountains—through snowy Vermont and Massachusetts—to Boston. Then the winter snow melted, and the cannons got stuck in thick mud. What a mess! Henry Knox was one of those people who never give up. Somehow, he got the cannons to Boston.

It was 1776, and when the British saw those guns in place, they pulled out of Boston. They took many of Boston's Royalist-Loyalist families with them. The Fluckers left Boston because of Henry Knox's cannons.

Henry Knox became one of George Washington's best friends. (That surprised Washington. He didn't think he'd like the New England Yankees until he got to know Henry Knox.) Knox and his artillery served in almost every major battle of the Revolution. After the Revolution, Washington made him the new nation's first secretary of war. That meant he was in charge of the army and navy.

Henry Knox loved to entertain and he loved to eat. He died after swallowing a chicken bone. It stuck in his intestine.

Mary Katherine Goddard would never have swallowed a chicken bone. She was the kind of person who did everything carefully and well. She had to. She had a brother with a hot temper who always seemed to be getting her in trouble. Mary Katherine was the steady, dependable one in the family.

She, too, had lost her father at an early age; she and her mother and brother learned the printing trade.

Printing was hard work in those days. Metal letters had to be arranged in the right order and set into a frame. Next they had to be inked. Then each sheet of paper was pressed onto the inked plate. It was slow going. Mary Katherine became an expert printer.

Because of her wild brother, she moved from city to city. That's how she got to Baltimore, where she ran a newspaper and printing business. It was Mary Katherine Goddard who had the courage to print the first copies of the Declaration of Independence, the ones signed by the members of the Second Continental Congress. She was lucky the British didn't close her printing press or put her in jail. They probably would have, if they hadn't had so many other things to worry about in 1776.

That same year she printed a satirical article signed "Tom Tell-Truth." The article said the Americans should do everything the British asked them to do. Satire is a form of humor. It often uses irony, which means saying one thing when you mean the opposite.

Print shops like Mary Katherine Goddard's were very important centers during the Revolution. Through them the politicians and agitators could get their message to as many people as possible.

(Ben Franklin loved satire. When he was in England, he wrote a pretend-serious article about whales that jumped up Niagara Falls. Some English people actually believed him. He was trying to show them how ignorant they were of America.)

Well, Tom Tell-Truth was trying to be funny, but some Patriots didn't have a sense of humor. A mob wanted to know who Tom Tell-Truth was. They wanted to close Mary Katherine Goddard's paper. She wouldn't tell them who wrote the article. Finally, the governor spoke out on her behalf. Today journalists still insist on the right not to reveal the source of their stories.

Mary Katherine Goddard became postmistress of Baltimore—until her job was given to a man. Two hundred and thirty businessmen in Baltimore signed a petition—they wanted her as postmistress. They said she was outstanding. But she didn't get the job back.

Maybe that got her thinking about unfairness. Maybe that was why Goddard freed her slave, Belinda Starling. In her will, Goddard left all her property to Starling.

In case you're curious, I can tell you now who Tom Tell-Truth was. He was Samuel Chase, a signer of the Declaration of Independence.

30 When It's Over, Shout Hooray

General Washington, in skirts, beats Britannia. His whip has 13 lashes, one for each colony.

Now, back to the War of Independence.

The British had more fighting men, more guns, and more experience. But the Americans had a big advantage: they believed in their cause. In England the war was not popular, and the longer it lasted, the more unpopular it became. It went on and on and on—for more than eight years. Besides, the military leaders in England were trying to plan a war that was being fought thousands of miles away. That never works well.

After the American victory at Saratoga, the war in the North became stalemated. That means it was even. That was good for the Patriots. Holding on was a kind of victory for the Americans; the British had to beat the rebel forces in order to win. So the English generals tried a new strategy: they shifted the war south.

The American Revolutionary War lasted almost nine years and was longer than any war in American history until the Vietnam War in the 20th century. It actually went on for two years after the battle of Yorktown, but mostly there were just small skirmishes. The battle of Yorktown convinced most people—but not King George III—that Great Britain had lost.

By 1778, three years into the war, Sir William Howe had gotten tired of the war and of being criticized for the way he was running things, so he resigned. General Henry Clinton became the new commander in chief of the British forces. Clinton believed the South was full of Loyalists and that they would help the English soldiers. He named Lord Charles Cornwallis commander of his troops in the southern states. Then he loaded soldiers onto ships

Lord Cornwallis won battles in the South, but he lost many men. "What is our plan?" he wrote. "Without one, we cannot succeed, and I assure you I am quite tired of marching about the country."

in New York harbor and sent them south. (Clinton kept a force in New York to hold on to that important city.)

Cornwallis was an able leader. First the British captured Savannah, Georgia. A British colonel wrote of ripping "one star and one stripe from the Rebel flag of America." He was talking about Georgia. It seemed to be in British hands. Next Cornwallis took Charleston, South Carolina. An American who was there described the British attack.

> *It appeared as if the stars were tumbling down...cannon balls whizzing and shells hissing continually amongst us; ammunition chests blowing up, great guns bursting and wounded men groaning.*

The British won again at Camden, South Carolina. That was a big win.

The siege of Yorktown really succeeded because the French navy drove off the British rescue fleet sent from New York. After that, Cornwallis and his redcoats were trapped.

England thought it had won the South, but those who believed in the Patriot cause wouldn't let them have it. Americans formed guerrilla bands and fought as the Indians did—with raiding parties. "We fight, get beat, rise, and fight again," said Nathanael Greene (the same man who was quartermaster general at Valley Forge). It must have been frustrating for the English officers. They kept winning the big battles, but they seemed to be losing the war.

Then came the most important battle of all, the battle of Yorktown.

Yorktown is a river port, near the Chesapeake Bay in Virginia. That's where General Cornwallis brought his troops in August 1781. It seemed an ideal headquarters spot for an army that got its supplies and support from the sea. (This is a good time to check a map.) Cornwallis's boss, General Clinton, was at the British military headquarters in New York; Clinton promised to send men and supplies by sea. The British were sure they would soon control Virginia.

Washington and a French general, the Comte de Rochambeau (kont-duh-ROSH-um-bo), were in Rhode Island making plans. At first they thought they would march their armies to New York, although they knew that city would be hard to take. Then they got word that a French admiral, Admiral de Grasse, was sailing from Haiti in the West Indies to Chesapeake Bay with a fleet of 28 ships. Could he blockade the bay and keep supplies from Cornwallis? That was what they hoped would happen. Rochambeau and Washington decided it was the chance they had been waiting for. They knew they would have to march their troops south—almost 500 miles. They had only a few weeks to do it; the French fleet couldn't stay for long.

The American war is over, but this is far from being the case with the American Revolution. Nothing but the first act of the drama is closed.

—Benjamin Rush

After the war, some 100,000 Loyalists moved from the United States to Canada.

John Paul Jones

John Paul Jones was a Scottish-born merchant seaman who became America's first naval hero. In 1779, his ship, the *Bonhomme Richard* (the French name for Ben Franklin's Poor Richard), sank a British warship, the *Serapis*. The English called him an outlaw and portrayed him as a devilish pirate.

They marched south together, and it must have been some sight. The French officers were elegant in white uniforms with gold braid. Their horses pulled wagons holding chests full of coins.

Most of the American officers wore bright blue uniforms with cream-colored trim (called buff). By this time many American privates (the ordinary soldiers) had uniforms, too, although they were often torn and ragged. But it didn't matter; the soldiers marched proudly with their general. They had become a disciplined army.

At Yorktown, three great military leaders greeted them: the dashing Frenchman, the Marquis de Lafayette; the cheerful German, Baron von Steuben; and a bold American, General Anthony Wayne (who was called "Mad Anthony" because he was so daring). They had great news for General Washington.

The French admiral, the Comte de Grasse, had arrived at Chesapeake Bay, fought the English fleet, and sent it sailing back to New York. And that wasn't all. De Grasse had brought extra troops who could fight on land. When George Washington heard all that news he took off his hat and handkerchief and waved them about. That was unusual behavior for the dignified general. "I have never seen a man moved by a greater or more sincere joy than was General Washington," wrote a French duke. When a French general stepped ashore, Washington gave the startled officer a big hug.

After beating off the British fleet, the French admiral de Grasse sent ships to fetch the American troops to Williamsburg.

The French-American army moved into Yorktown. They dug deep trenches at night. In the morning the British redcoats found themselves trapped. A half-circle of entrenched soldiers faced them. The York River was behind them. The Americans began firing their cannons. Then a brave young colonel named Alexander Hamilton led an attack. He captured a key British earth fortress.

The British didn't have a chance. They were outnumbered and outflanked. Cornwallis did everything he could. He even tried to save his army by sailing his soldiers across the York River to safety. But he had bad luck—a sudden storm swamped the boats.

When the British army surrendered at Yorktown on September 18, 1781, it was exactly four years to the day after Gentleman Johnny Burgoyne's surrender at Saratoga.

On October 19, 1781, the British surrendered at Yorktown. Lord Cornwallis could not bring himself to hand over his sword in person, so Brigadier-General Charles O'Hara of the Guards did the deed. The War of Independence was over.

In this drawing scratched on a powder horn, a band of Continental soldiers moves a siege cannon toward enemy lines. This type of cannon was used at Yorktown.

The British adventure in America was coming to an end at Yorktown, just 25 miles from Jamestown, where it had all begun.

An English drummer boy climbed on top of a trench and beat his drums. An officer followed waving a white handkerchief. The great British army was surrendering. It was October 17, 1781.

Two days later, American soldiers stood proudly in a long line; facing them was a line of happy French soldiers. Between them marched the British and German armies; the defeated men were wearing clean uniforms and trying to keep their heads high, but many British soldiers cried when they laid down their arms. Army bands played an old English nursery tune, "The World Turned Upside Down." Here are the words and music:

If buttercups buzzed after the bee;
If boats were on land, churches on sea;
If ponies rode men and grass ate the cows;
And cats should be chased to holes by the mouse;
If the mammas sold their babies to the gypsies for half a crown;
Summer were spring and the t'other way round;
Then all the world would be upside down.

And upside down it was. David had licked Goliath. The colonies would soon be states; the infant New World was growing up. A superpower had been defeated by an upstart colony.

A new nation was being formed: a nation that would try not to make the mistakes of its European parents. A nation that would be founded on ideas of freedom and equality. A nation ruled by laws, not kings. That nation soon had a great seal—which you can see on every dollar bill. On one side are two Latin words, *annuit coeptis*—"[God] has favored our venture." On the other side are the Latin words *novus ordo seclorum*. They mean, "A new order of the ages [is created]."

The official end of the war came on January 14, 1784, when Congress ratified a treaty signed in Paris the year before.

31 Experimenting With a Nation

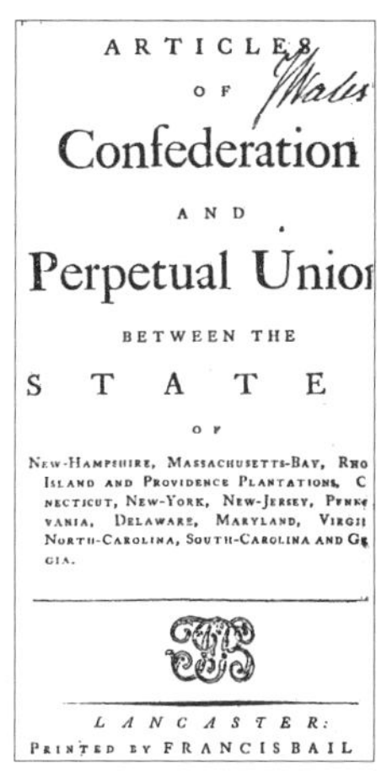
ARTICLES
OF
Confederation
AND
Perpetual Unio
BETWEEN THE
S T A T E
OF
New-Hampshire, Massachusetts-Bay, Rho Island and Providence Plantations, C necticut, New-York, New-Jersey, Penns vania, Delaware, Maryland, Virgi North-Carolina, South-Carolina and G gia.

LANCASTER:
Printed by FRANCIS BAIL

The Articles of Confederation were the country's first constitution—but they were too weak to do a good job.

Imagine a city built of wooden blocks. Do you see it in your mind? Make sure it has houses and bridges and walls. Knock it down. Now build it again.

Which takes longer, destroying or building? Which is harder?

It's the same way with governments.

Revolutions are difficult—overthrowing Britain wasn't easy at all for the American colonists—but building a strong nation was much harder.

The American Revolution was unusual; it produced people who were good at nation-building. When you study other revolutions, like the ones in France and Russia, you'll see how lucky we were.

At first, though, it looked like it might not happen. It seemed as if the 13 states would never get along. They certainly weren't "united." Each state was printing its own money and making its own rules. Eleven states had their own navies. Virginia's navy had 72 ships. The Continental Congress was trying to run a national government, and it had a navy, too—but it was smaller than Virginia's. The Congress was also printing money. As you can guess, soon none of the money was worth anything, and that was terrible for most citizens.

Besides all that, each state got into the taxing business: New York was taxing goods from New Jersey, and New Jersey was taxing goods from New York. Virginia and Maryland were

In 1782, Colonel Lewis Nicola wrote a letter to General Washington suggesting that he use his army to seize power and proclaim himself king. Washington replied, "You could not have found a person to whom your schemes were more disagreeable."

Supply & Demand

There is an economic law called the law of *supply and demand.* If there is a big supply of something, the price—and the demand for it—usually goes down. Gold is expensive because it is beautiful and *rare.* If there were gold nuggets all over the place, the price of gold would go way down. Money works in roughly the same way. If a government prints lots of money, the value of its money goes down. That means it costs many dollars to buy something that once took only a few dollars. That is called *inflation.*

Everyone who could make a claim to the lands west of the Appalachians and east of the Mississippi was doing it. Some areas were claimed by three or four states at once. (See key on the facing page.)

squabbling over boundary lines. Little states were jealous of big states—and vice versa. In Massachusetts some farmers rebelled against the government in Boston. In Philadelphia and New York newspapers reported a movement to create three separate nations out of the 13 former colonies. In England people were saying that the Americans would soon be begging to be taken back.

As you can see, the United States got off to a rocky start. We didn't have a good working plan for a government.

We didn't begin with the Constitution we now have. The first constitution of the United States was called the Articles of Confederation. It didn't work well at all.

George Washington, in a letter to the Marquis de Lafayette on May 10, 1786, talked of our "constitution" and "President." He was referring to our first constitution, the Articles of Confederation, and to President John Hanson.

That was because the American citizens were afraid of political power. They had had a bad experience with kings and parliament. They were afraid of a strong congress and of a strong president. So they went to the other extreme. They didn't give Congress the power to do much of anything. There was no president except the president of the Congress. And there wasn't much he could do either.

Ask anyone, "Who was the first president of our country?" The answer will be "George Washington." But you can say that the first president was John Hanson. Very few people will believe you. It's true, though. Hanson became president under the Articles of Confederation,

CONN.

N.H.

MD.

N.Y.

VA.

S.C.
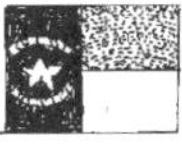
N.C.

MASS.

PA.

R.I.

N.J.

on November 5, 1781. President Hanson didn't make himself remembered, because he had no power.

In 1781 Americans were facing one of the toughest problems there can be in designing a government. How do you provide freedom for each person and still have a government powerful enough to accomplish things?

The Articles of Confederation were written by a committee appointed by the Second Continental Congress on July 12, 1776. They were ratified in 1781 and lasted until 1789.

You have to give up some freedom when you are part of a society that is ruled by laws. The question is, how much do you have to give up? The Americans, at the end of the 18th century, had just fought hard for liberty. They weren't about to give up much at all. They went too far—but they learned.

The national government, under the Articles of Confederation, was just too weak. Everyone seemed to know it. Most of the time the states wouldn't even send representatives to Philadelphia to vote at meetings of the Congress. A lot of people felt the voting wasn't fair anyway. Each state had an equal vote in congress. That meant that 68,000 Rhode Islanders had one vote, and so did 747,000 Virginians.

Then something really insulting happened. In 1783 Congress got chased out of Philadelphia by its own army, because it hadn't paid the soldiers their salaries. But Congress had no money to pay the salaries and no power to collect taxes. (It is tax money that governments use to pay their bills.)

In 1787, when this cartoon was printed, America's government was in a mess. Two groups of six states are engaged in a tug-of-war, each trying to pull a wagon labeled "Connecticut" over to their side.

It would take six years for the people living in this land to create a workable kind of government.

At first the former colonists didn't even know what to call themselves. We began as a nation without a name. Some called us the American Commonwealth; others said the American Confederation. Some talked of "united states"; a few said *the* United States.

But most people still thought of themselves first as citizens of the state they lived in. They were having a hard time accepting the idea of a nation that might be more important and powerful than their separate and beloved states. In

England's Indian Ally

Mohawk Joseph Brant (you remember, he was William Johnson's brother-in-law) was now fighting settlers in western New York and Pennsylvania. Easterners were ignoring Indian treaties and moving into those regions. The English were secretly helping Brant, who was a skilled warrior. They didn't think the new nation would last long. In 1785 Brant went to England and met George III. He made a big hit in England. Brant was highly educated and had translated the Bible into Mohawk.

fact, they didn't even like the word "nation." They called it a "union" of states.

People in the territories felt the same way. You already know about independent Vermont. Well, some people tried to make Kentucky into a nation, too. (In 1792 the Commonwealth of Kentucky became the first state west of the Appalachian mountains.) There was even a state that called itself Franklin, off to the west of North Carolina. It was territory where Mound Builders had once flourished. Before long Franklin became a state with an Indian name: Tennessee.

Settlers were filling up the Ohio River Valley, and that was causing problems. Much of that western land was claimed by Virginia, but other big states were claiming some of it too. The states without western lands were jealous. How could arguments between the states be settled unless a central government had more power than any one state?

There was one good thing about the Articles of Confederation: they were so weak they made a strong constitution possible.

In this British cartoon, a triumphant America has laid down bow and arrow and is offering the olive branch of peace to a weeping Britannia.

32 Looking Northwest

A wagon train musters for the westward trail. The wagons took the name of the Pennsylvania valley where they were first built: Conestoga.

Maryland was being difficult, and so was Rhode Island. They just weren't going to let big Virginia hog everything. They certainly weren't going to cooperate as long as the large states had western land and they didn't. So they finally got Virginia and Georgia and some other big states to cede—or give up—all claims to western lands. It was agreed that those western lands—if they ever had enough people—would also become states.

Then, in 1787, the Confederation Congress passed the Northwest Ordinance. (An ordinance is a law passed by a government.) If you got the impression that the Congress under the Articles of Confederation was a total washout, that isn't quite true. That congress did a few things right, and the Northwest Ordinance was one of them. It provided a fair way for new territories to become states. It was another American "first" in world history.

Virginia gave up enough land in the Northwest Territory to make the future states of Ohio, Illinois, Indiana, Michigan, Wisconsin, and part of Minnesota. Without Virginia to claim it, that land was like a colony belonging to the 13 states. Now colonies throughout history had always existed for the benefit of the mother country. The Americans certainly knew all about that! Because of their own bad experience with Great Britain, they didn't want to take advantage of others. So the Northwest Ordinance was based on equal rights for the territories. It was based on fairness.

Spain Again

Spaniards were encouraging Indian tribes to harass the settlers in Tennessee and Georgia and Kentucky. Spain controlled Florida, and she still had her eyes on Georgia, next door. And no one knew who would end up with the lands across the Appalachians—the Spaniards? the English? the Native Americans? or the new republic of American states? Spain controlled the Mississippi and the important port of New Orleans, and people in Tennessee and Kentucky shipped goods down that big river. They needed to get along with the Spaniards.

This 1783 map recorded the boundaries of the new United States. Most of the western territory was still Indian land—but the Indians were soon pushed farther and farther back.

A system was devised for dividing the land into areas called townships. Groups of townships could become states. That system worked so well it was used again and again as the nation grew.

Great Britain had tried to keep settlers out of the western territories. It had tried to keep those lands for the Indians. It was easier to rule that way. Now, with Great Britain out of the picture, settlers were heading west. Once again, the Native Americans were going to be pushed from their lands. Disease and guns killed many; some joined white society; most fled farther west.

No one knew how many settlers were moving over the Appalachians—the first census did not come until 1790—but thousands and thousands were on their way west. Many piled their families and their belongings into big wooden-wheeled wagons and hitched them to oxen. They were called covered wagons because they had canvas tops that were stretched over curved wooden rods; that made each wagon into a big room on wheels.

For a definition of habeas corpus, *see chapter 10.*

The people who moved into the Northwest Territory were guaranteed freedom of religion, habeas corpus, and trial by jury because the Northwest Ordinance had a bill of rights. That ordinance provided another very important guarantee. Here it is—pay attention to this one—"there shall be neither slavery nor involuntary servitude in the said territory."

Involuntary servitude means having to work for someone whether you like it or not. That was what slaves did, and indentured servants too. The Northwest Ordinance outlawed both.

The Northwest Ordinance also said, "Religion, morality and knowledge, being necessary to good government and the happiness of mankind, schools and the means of education shall forever be encouraged."

If people were to govern themselves—as Americans were doing—then they had to be educated. How can you govern yourself if you can't read or write? How can you take part in government if you don't know about current events—and history, too? What should schools be like? What should they teach? And should they be for everyone?

These were questions that Americans were asking at the end of the 18th century. The old-style rulers had tried to make sure that most of

their subjects were not educated. They knew that reading gives you power. If people can't read, then they don't know what is going on, and that makes them depend on a ruler. Thomas Jefferson believed that you couldn't have a people's government if the people were ignorant. This is what he wrote in a letter to a friend: "If a nation expects to be ignorant and free, in a state of civilization, it expects what never was and never will be."

The Northwest Ordinance required that each township set aside land for public schools—and that was Thomas Jefferson's idea. Finally, the ordinance said, "The utmost good faith shall always be observed towards the Indians." What do you think of that statement? Do you think it was followed?

A pioneer family rests on the road to Pittsburgh. Their Conestoga is piled high with feather beds, to be pulled out at night and spread under the wagon.

33 A Man With Ideas

An Italian Immigrant

When Thomas Jefferson felt like a little mental exercise, he could climb on his horse and gallop over to see Philip Mazzei at his nearby estate. Mazzei had moved to Virginia from Italy. Like Jefferson, he had a wide range of interests. Mazzei was fascinated with growing things, and he introduced the cultivation of grapes, olives, and some other fruits into Virginia's Piedmont. (Winemaking is an industry of the area today.) Mazzei was a surgeon, a wine merchant, and a writer. In 1775 he wrote an article in Italian that was translated by Thomas Jefferson and printed in the *Virginia Gazette*. It said, "All men are by nature equal, free and independent." Jefferson was so impressed by Mazzei that he wrote to his friend Richard Henry Lee that Virginia would be much more prosperous if the colonists had been Italian instead of English.

Jefferson in Paris. In 1783 he took over from Franklin as ambassador to France.

This is a good time to stop and tell you about Thomas Jefferson. The more you learn about United States history, the more you will hear his name. You will find that some people describe him as an architect; some call him a scientist; others say he was a great writer; and still others talk of him as a politician.

Well, he was all of those things, and more, too. But mostly he was an intellectual: an idea-centered person. America is a country built on ideas, and it was Thomas Jefferson who wrote down many of our best ideas with his quill pen.

The words of Jefferson's Declaration of Independence made the American Revolution more than just a war. It became a war for an idea. Jefferson thought up most of the ideas in the Northwest Ordinance. Today, when people talk about human rights, they usually quote Thomas Jefferson.

Jefferson was an inventor, a scientist, an architect, a surveyor, a landscape architect, a lawyer, a farmer, a political philosopher, a statesman, and a superb horseback rider. He loved beautiful things, good food, and

Jefferson at Virginia's House of Burgesses. He was elected at the age of 25.

Jefferson made this sketch of his idea for a house in 1771. When Monticello was eventually built, the main part of the second floor had become an octagon (it had eight sides).

good conversation. He played the violin. He had a fine voice, and when he was on horseback he usually sang. It was Jefferson who insisted on a money system for the United States based on tens—the decimal system. English money was not based on tens, and was hard to figure.

He was 6'2" tall, with the kind of frame that seems to be all bones. His face was long; his eyes were thoughtful; his hair was red. People liked Jefferson, but his personality was not like that of the fun-loving Benjamin Franklin. He was more serious. Franklin and Jefferson were similar in some ways, however. Each had a cheerful, optimistic personality and each saw the best in people. All their lives they were question-askers, always experimenting and doing. Jefferson founded the University of Virginia and designed its buildings around a great central lawn. Many architects consider it the finest campus plan in the world.

If you look at a nickel, you will find Jefferson's head on one side and his elegant house on the other. He called the house Monticello, which is Italian for "little mountain." He spent much of his life building that house. Inside are doors that seem to swing by magic—Jefferson designed them. He filled Monticello with collections of things that he found interesting, like books, animal bones, and Indian artifacts. He stocked the pond with fish and the gardens with unusual flowers, fruits, vines, and trees.

Jefferson was born and grew up in a part of Virginia known as the Piedmont region. In French, *pied* means "foot" and *mont* means "mountain." Foothills, at the base of high mountains, are called *piedmont*.

When Tom was a small boy in the Virginia Piedmont, there were more wolves and bears and deer there than people. Later, no matter where Jefferson lived—Williamsburg, or Philadelphia, or Paris, or Washington—he would always be homesick for Virginia's beautiful mountains.

His father, Peter Jefferson, was a strong-muscled, fair-dealing man

Reason to Rejoice

A letter to members of the New Church in Baltimore, 1793:

We have abundant reason to rejoice that in this Land the light of truth and reason has triumphed over the power of bigotry and suspicion, and that every person may here worship God according to the dictates of his own heart....In this Land of equal liberty it is our boast, that a man's religious tenets will not...deprive him of the right of attaining and holding the highest Offices that are known in the United States.

—George Washington

who was at home on the frontier and also in the parlors of the Virginia gentry. Thomas Jefferson once wrote to John Adams:

> *In the very early part of my life I was very familiar [with Indians]... Before the Revolution they were in the habit of coming often, and in great numbers to [Williamsburg]...I knew much of the great Outassete, the warrior and orator of the Cherokees. He was always the guest of my father, on his journeys to and from Williamsburg.*

Peter Jefferson's "education had been quite neglected," his son wrote, but "he read much and improved himself." Perhaps because he had little schooling himself, he valued education and made sure his son had the best teachers he could find. Peter was a surveyor who drew a famous map of Virginia and taught surveying skills to his son. He died when Tom was 14. Jefferson's mother, Jane Randolph, was a member of a wealthy family that owned large plantations and many slaves.

Jefferson was lucky. His parents and teachers encouraged him to use his mind—and he did. When he was 16 he enrolled in the College of William and Mary in Williamsburg. There he was really lucky. He was befriended by three important people: the best teacher at the college (William Small); the best lawyer in Virginia (George Wythe—say *with*); and the royal governor (Francis Fauquier—say FAU-keer). They must have realized that he was an unusual young man, because he was often invited to dine and talk and play the violin with them. It was George Wythe who taught Jefferson law. Later he wrote that whenever he was tempted to do something he knew he shouldn't do, "I would ask myself what would Dr. Small [or] Mr. Wythe...do in this situation?"

Jefferson became Virginia's governor, the United States ambassador to France, secretary of state, and president. He didn't seem to think any of those accomplishments were as important as the three things he asked to have listed on his tombstone. They were:

- *that he was author of the Declaration of Independence*
- *that he wrote the Virginia Statute for Religious Freedom*
- *that he was Father of the University of Virginia.*

Jefferson's dumbwaiter, hidden inside a fireplace, brought up wine bottles from the cellar.

Cow Cure

An English physician, Dr. Edward Jenner, noticed that milkmaids never caught smallpox. He knew that cows got a mild form of the disease. Could milkmaids be catching cowpox—and was it that that protected them from the deadly version? Jenner made a serum from cowpox pus and injected it in his young son. It worked. But most people were too scared to try this new idea. What would you have done? James Franklin (Ben's older brother) wrote newspaper articles telling of the dangers of vaccination. (Some people did die.) Thomas Jefferson bravely tried the new procedure. And Ben Franklin believed in it, too. But his beloved four-year-old son Francis hadn't been inoculated when an epidemic swept through Philadelphia. Francis died of smallpox. Experts say that 36 million people may have died of smallpox in the 18th century. Today, thanks to Dr. Jenner, we don't have to worry about smallpox.

Jefferson also invented this polygraph, which made a copy of a letter with one pen as you wrote with the other.

That Statute for Religious Freedom did something that had never been done before in all of history. It said—officially—that governments have no business telling their citizens what to believe. In other words, a government and its citizens' religions should be separated. That is known as *separation of church and state*. It is the foundation on which religious freedom is based. Some say it is the basis for all other freedoms.

Jefferson called it freedom of conscience. He meant for people to be free to believe whatever they wished. The power of government, he said, should only extend to acts that hurt others. "But it does me no injury for my neighbor to say there are twenty gods or no god. It neither picks my pocket, nor breaks my leg."

That was a daring idea in the 18th century. It was something to gasp at. No other state in the world allowed its citizens complete freedom to choose their own religion—or to choose not to have any religion at all.

It took nine years for Jefferson and his friend James Madison to convince the Virginia legislature to pass the Statute for Religious Freedom. Jefferson said that getting the statute passed was "the severest contest in which I have ever been engaged." That means that nothing he ever did was as difficult. It was worth all the hard work. American thinkers were doing things no political leaders had ever done before.

The word ***state*** is often used to mean nation or government.

They Called It Macaroni

When Jefferson came home from Europe, he tried to bring good European ideas with him. He brought ideas on architecture and used them to design the Virginia Capitol in Richmond. He brought ideas from books and shared them with his good friend James Madison. And he brought new foods and served them at his dinner table at Monticello.

But he didn't have a pasta-making machine. So he wrote his secretary, William Short, who had stayed behind in Europe, and asked him to get one. Pasta is an Italian specialty. Short went off to Naples in Italy to get Jefferson a machine. He called it a "macaroni mould."

This is how Short described it in a letter to Jefferson: "It is of a smaller diameter than that used at the manufactories of macaroni, but of the same diameter with others that have been sent to gentlemen in other countries. I went to see them made. I observed that the macaroni most esteemed at Naples was smaller than that generally seen at Paris. This is the part of Italy most famous for the excellence of the article." Short didn't know he was describing spaghetti.

Some people in the colonies didn't like all these fancy foods that Jefferson brought from Europe. Blustery Patrick Henry was one of them. He liked to stick to the old ways. The foods his father had eaten were good enough for him. He said Jefferson was one who "abjured his native victuals." In today's English that means someone who won't eat his home food. (Do you think that was true?)

34 A Philadelphia Welcome

At the Constitutional Convention, Edmund Randolph hoped for "zealous attention to the present American crisis."

Everyone agreed that Philadelphia was the most modern city in America, perhaps in the world. Boston's narrow, twisting streets reminded people of Europe's cities. But Philadelphia had straight, broad avenues that crossed each other making nice, even, rectangular blocks. Water pumps were spaced regularly on each block, and street lamps —662 of them—lit the city at night. Philadelphia's streets were paved with cobblestones or brick. Some people complained of the noise when horses' hoofs clattered over the stones, but they had to admit it was better than having dirt roads that turned dusty or muddy with the weather. Philadelphia even had sidewalks. They were edged with posts to protect pedestrians from the traffic.

***Accident on Lombard Street* in Philadelphia, 1787:**

*The pye from Bakehouse she
had brought
But let it fall from want of
thought
And laughing sweeps collect
around
The pye that's scatter'd on the
ground.*

Horse and carriage traffic was heavy, and accidents were commonplace. It was to be expected: Philadelphia, with 40,000 people, was the largest city in North America. It was a city proud of itself. After all, it had 7,000 houses, 33 churches, 10 newspapers, two theaters, a university, a museum, and a model jail.

The jail was across a lawn from the imposing State House. The language that came out of the windows of the jail was not model English. Ladies covered their ears and hurried by. At the red-brick State House, the language was courtly and proper. Nearby, at the new home of the Philosophical Society, the language was scholarly. There, great men, like the famous Dr. Benjamin Franklin, discussed science and the latest ideas.

Despite their dignified Quaker beginnings, Philadelphians loved parades and celebrations. So when George Washington rode into town on May 13, 1787, to attend the convention that was to write a new constitution for the new nation, it seemed as if all 40,000 people came out to cheer. Church bells rang, cannons were fired, and those who lined the streets applauded the great general who had done so much to make the country free.

James Madison was one of the first delegates to arrive in Philadelphia, but no one paid him attention. He was 11 days early for the Constitutional Convention, and he was not a celebrity. Madison came by coach from New York, and was sore and tired after being squeezed in with a dozen others on hard, backless benches. The coach, pulled by horses, took two days to make the trip. It was called the *Philadelphia Flier*.

Madison checked into Mrs. Mary House's boardinghouse: it was quiet and convenient and less expensive than the popular inns. There he could work hard without interruptions.

Those who knew Madison weren't surprised that he was early. They said that was typical of him. He liked to be prepared.

Some Philadelphians thought all Virginians were giants—until they saw James Madison. He was small and soft-voiced. Someone once described him as "no bigger than a half piece of soap." But he was well put together, in mind as well as body.

His eyes were blue as a May sky. He had a boy's look and seemed even younger than his 36 years. Perhaps that was why he wore black suits and pulled his hair back and powdered it white in a style that made young men seem old and wise. People liked James Madison; his quiet, sensible ways impressed them. You could tell right away that he was a thinking man. His friends called him Jemmy.

Jemmy Madison was the oldest of 12 children born to a plantation-owning Virginia Piedmont family. While most Virginians went to the College of William and Mary, Madison chose to go north to Princeton, which was then called the College of New Jersey. That gave him ideas and friends he might not have had if he had stayed close to home.

At first he thought he would be a minister of the church. Then he changed his mind and studied law; he said it was so that he could "depend as little as possible on the labor of slaves." It turned out that he didn't like reading books of law. He gave that up and spent the rest of his life as a political leader. It was the labor of

James Madison was 27 years old when he helped write Virginia's constitution.

Philadelphia's Walnut Street Prison was toured by visitors from Europe, who came to admire its humane conditions.

slaves that allowed him the freedom to do what he wished. Madison hated slavery, but he didn't know what to do about it.

More than anyone else, it was Madison who got this convention organized. He wrote letters to Washington, to Jefferson, to Adams, and to others, urging them to attend. The convention was supposed to revise the Articles of Confederation, but Madison thought rewriting it a poor idea. He believed the Articles should be scrapped—thrown out—and a whole new constitution written. He knew he would have to convince a lot of delegates of that, so he went about it the way he knew best—by studying.

Madison was a scholar. He read all he could find about governments all over the world and throughout history. Long before the convention got started, he wrote to his good friend Thomas Jefferson and asked for help. Jefferson had taken Franklin's place as America's minister in France. Jefferson sent Madison books—hundreds of books—and he sent his ideas.

Madison read about the governments of ancient Greece and Rome and of other places and times. Then he took the best ideas he found and wrote them in notebooks that he brought with him to the convention.

By this time, the other delegates were in Philadelphia and ready to get started. Madison and the Virginians set to work discussing a new plan of government that Madison had written. They agreed to have Virginia's popular governor, Edmund Randolph, present the plan to the Constitutional Convention. It was called the Virginia Plan, and it made things much easier for all the delegates at the convention. When they began their meeting, they had a document in front of them. It gave them a starting point; it helped speed up the process.

And that summer anything that made work easier was appreciated. Some people said that 1787 was the hottest summer in Philadelphia's history. They were exaggerating. It was probably no hotter than usual in Philadelphia that summer—but it was hot.

Philadelphians shopped for groceries at the city's Country Market Place.

James Madison

This is how William Pierce, a Georgia delegate, described Madison:

Mr. Madison is a character who has long been in public life; and what is very remarkable is that every person seems to acknowledge his greatness. He blends together the profound politician with the scholar...and though he cannot be called an orator, he is a most agreeable, eloquent, and convincing speaker. From a spirit of industry and application which he possesses in a most eminent degree, he always comes forward the best informed man of any point in debate. The affairs of the United States, he perhaps, has the most correct knowledge of, of any man in the Union....He is easy and unreserved among his acquaintance, and has a most agreeable style of conversation.

A Philadelphia Gardener

This is the story of a gardener who knew Franklin and Washington and most of the people you've been reading about. He wasn't a politician—he was, as I said, a gardener. He became a famous scientist and America's first great naturalist (which means he loved and studied nature). And there is an old tale that some of his plants—sent to England—may have led to the Revolutionary War.

You see, the Prince of Wales (who was the king's oldest son, and expected to be king himself) was an amateur gardener. So when some rare trees arrived from America, the prince insisted on standing out in the rain until they were planted. He got wet, caught pneumonia, and died. His younger brother, who became king, was George III. But if the prince hadn't died, and if George III hadn't become king, who knows what might have happened between England and America?

Which makes John Bartram mighty important. He was a Quaker, born in 1699, and a founding member of the famous Philosophical Society. Bartram loved plants and animals and he wanted to travel and study America's wildlife, but he had a family to feed. Then some men in England heard about him and paid him to send them examples of flowers, birds, insects, and animals. So he became a botanist to the king and the Royal Society (a kind of academy of scientists and scholars). It was exciting for him, and for them, too. Once he shipped turtle eggs to London and they hatched the day they arrived. "Such a thing never happened, I daresay, in England before," wrote Peter Collinson, who watched the turtles hatch.

When Bartram went to Virginia to study plants, Collinson wrote telling him to be sure to get some new clothes.

> *For though I should not esteem thee the less, to come to me in what dress thou will—yet these Virginians are a very gentle, well-dressed people—and look, perhaps, more at a man's outside than his inside. For these and other reasons, pray go very clean, neat, and handsomely dressed to Virginia.*

One of the people he visited in Virginia was Isham Randolph, Jefferson's grandfather.

Bartram went off into the woods, where the mountain lions didn't scare him but some of the Indians did.

> *An Indian man met me and pulled off my hat in a great passion, and chawed it all around—I suppose to show me that they would eat me if I came to that country again.*

When he went to Florida, Bartram took his son William, who was an artist as well as a naturalist. William liked Florida so much he stayed for several years, drawing alligators, birds, and snakes as well as flowers and trees. William carried on his father's work, and Jefferson asked him to go on a scientific expedition across the country to the far ocean. (John Bartram had suggested that idea to Franklin, who had told Jefferson.) WIlliam was too old to make the trip, but he helped with the planning. That 1803 trip was the Lewis and Clark expedition. John Bartram's garden, full of rare plants, is now part of the Philadelphia park system.

William Bartram packed into one picture (below) the American lotus, the Venus flytrap (in lower left corner), and the great blue heron. Above, Franklinia alatamaha*—named for Ben Franklin.*

35 Summer in Philly

Washington didn't talk a great deal, but his calm presence kept things going.

It was already warm in late May 1787, when the Constitutional Convention was officially called to order. It was just beginning to cool down in September, when the Convention finally completed its business.

Besides the heat, there were flies and mosquitoes—big, biting flies and mosquitoes. They bit right through the silk stockings that the delegates wore.

In those days, no one knew much about sanitation; they didn't know those little pests could be dangerous. So people dropped their garbage in the streets. At night, pigs and cows were let out to eat up the garbage. They left their own droppings, and so did the horses, which filled the streets during the day.

There were no bathrooms then, just tiny outdoor rooms called "necessaries," or "privies,"

The Convention debates were endless. Washington (standing, right) said, "I see no end to my staying here."

At the end of the 18th century, the median age of Americans was 16. That means that half of all Americans were younger than 16 and half were older. Today the median age is about 33.

which were polite names for holes in the ground.

And people didn't take baths. They thought it unhealthy. When Elizabeth Drinker, a respectable Quaker lady, tried the shower her husband put in their backyard, she wrote in her diary, "I bore it better than I expected, not having been wet all over at once, for 28 years past."

As you might expect, Benjamin Franklin did have a bath. He never let other people's ideas stop him from experimenting. He had a round wooden tub built and set it on large paving stones that were heated by fires circulating hot air under the stones. The idea had been developed centuries earlier in China. Ben read about it, tried it, and liked it. Benjamin Rush, who was a member of the Philosophical Society and another advanced thinker, said that the heated bath "smoothed the descent of Dr. Franklin down the hill of life and helped prolong it beyond 84 years."

Ben Franklin by Charles Willson Peale, who painted many of the great men of the time.

Ben Franklin was always ready to try—or invent—new ways of doing things. The rest of Philadelphia smelled. Anyone with a cold or a stuffed nose was probably lucky.

It was heaven, though, for the flies and mosquitoes. They spread germs all over town, and that led to a lot of illness. In those days babies often didn't survive long. (Of James Madison's 11 brothers and sisters, five died as children. That wasn't unusual.) Mothers sometimes used tin nipples for their baby's bottles—they thought that was a good thing to do—but the poor infants got lead poisoning.

Grownups died young, too. If you went to a doctor he might decide to take some of your blood—sometimes a lot of it. Bleeding was supposed to be a cure for disease, but it could kill the patient. (Wormlike creatures, called leeches, were sometimes used to suck your blood.)

The 55 delegates to the convention all made it through the Philadelphia summer, so they must have been strong men. Many of them stayed at the Indian Queen, which was one of the finest taverns in America.

A visitor that summer described it as "kept in an elegant style, and consists of a large pile of buildings, with many spacious halls, and numerous...lodging rooms." A servant who took that traveler to his room was "a young, sprightly, well-built black fellow, neatly dressed... his shirt ruffled, and hair powdered...he brought two of the latest London magazines and laid them on the table. I ordered him to call a barber, furnish me with a bowl of water for washing, and to have tea

Behind Closed Doors

The Framers—the men at the Constitutional Convention—were deciding on a government for a free people. Why were they so secret about it?

***Reason 1*: they didn't want their constitution torn apart before it was even finished.**

***Reason 2*: the delegates wanted to be able to change their minds. At the Constitutional Convention, any subject—even if it had been decided—could be brought up and voted on again.**

Do you think those reasons are good enough? What are the advantages of private discussions? What are the dangers of closed lawmaking? Today reporters listen to congressional debates. So can you if you visit Washington, or watch cable TV.

It wasn't until after the Civil War that screens became popular. One Philadelphian, in Ben Franklin's time, put a hornet's nest in his dining room so the hornets would eat the flies and mosquitoes.

At taverns like the Indian Queen the customers passed the time with games such as billiards (below), which is similar to pool. Washington loved billiards, and also playing cards—though he was careful when he played for money.

on the table by the time I was dressed."

On the first floor of the Indian Queen, men sat at round tables in the public sitting rooms, drank toddies, lemonade, or Madeira wine, read newspapers, and exchanged gossip. (Women were not allowed in the public rooms.)

Not all the delegates could afford the Indian Queen, and some who did had to share rooms. Of course, there wasn't air-conditioning, or even screens. If they opened the windows the flies and mosquitoes came in; if they closed the windows it was too hot. No one got much sleep. Some days they must have been grumpy.

It was a good thing that James Madison was so well prepared. On top of everything else, he did something very important to those who would live after him: he took notes. That is how we know what happened.

You see, the convention voted to keep all its proceedings secret. George Washington enforced that rule. He had been elected president of the convention unanimously. When Washington was in charge, everyone did what he said.

Today our sessions of Congress are open to the public and the press. Democracy usually works best in an atmosphere of openness. Some people in 1787 didn't like the idea of secrecy. When Thomas Jefferson heard about it—remember, he was in France as ambassador—he was upset. But there are people who say the Convention never could have accomplished what it did if everyone in Philadelphia had known what was going on.

A secretary was hired to keep a record of the proceedings. His records were lost. But Madison sat close to the front, never missed a session, and copied down all the speeches. James Madison has been called the Father of the Constitution. However, he wasn't the only important man there. In the next chapter I will tell you about a man who slapped George Washington on the back—but only once.

36 A Slap on the Back

Let our government be like that of the solar system. Let the general government be like the sun and the states the planets, repelled yet attracted, and the whole moving regularly and harmoniously in several orbits.
—JOHN DICKINSON

Inside Independence Hall, the debates were secret. William Paterson reported rumors that the delegates were "afraid of the very Windows, and have a Man planted under them to prevent Secrets and Doings from flying out."

Benjamin Franklin was 81, the oldest delegate at the Constitutional Convention. Back in 1776, when he left England and sailed for Philadelphia, he thought his political career was behind him. He hadn't realized then—no one did—that some astonishing years lay ahead.

Now, 11 years later, Georgia's delegate, William Pierce, described him as "a most extraordinary man, and tells a story in a style more engaging than anything I ever heard. ...He possesses an activity of mind equal to a youth of twenty-five years of age." Unfortunately, Franklin's body had not stayed as young as his mind. He was in pain much of the time, from gout and from a stone in his bladder. Franklin could hardly walk, nor could he endure the shaking of a horse-drawn carriage. So he was carried—by strong convicts—in a sedan chair balanced on long poles. He'd brought the chair from Paris; he and the chair were among the sights to see that summer in Philadelphia.

If you were really fortunate you might get invited to dinner in the new dining room that Dr. Franklin had added to his house. It could seat 24 people. Ben wasn't the only Philadelphian who enjoyed company. It was a sociable city, and the delegates were special guests. Every host and hostess wanted General Washington as a guest of honor. Samuel and Elizabeth Powel invited him often. The Powels lived in an elegant family enclave in the middle of the city with a formal garden that connected four Powel homes. In his diary, George

Washington found some time to relax at the Convention, and asked that extra clothes be sent to him from Mount Vernon. This is how he dressed for an evening party.

Ben Franklin Makes Fun of Slave Owners

Benjamin Franklin had a talent for making bad ideas look foolish. A senator from Georgia gave a speech in the new Congress. He said that slavery was really good for the slaves and that slave owners were just following the Christian Bible. Franklin didn't believe that for a minute, but he knew that many people never questioned what their leaders told them. So he wrote a parody—a humorous imitation—in which he took the senator's ideas and turned them around. His parody was a made-up letter from a Muslim prince who ruled a nation where captive white Christians were used as slaves. That, said Franklin's pretend Muslim prince, was good for the Christians. Besides, he said, the slave owners were just following the ideas of the Muslim holy book, the Koran. Franklin put the senator's words in the mouth of Sidi Mehemet Ibrahim, his ruler from the African nation of Algiers. Of the slaves, Ibrahim–Franklin said:

"Is their condition then made worse by their falling into our hands? No; they have only exchanged one slavery for another, and I may say a better; for here ...they have an opportunity of making themselves acquainted with the true doctrine, and thereby saving their immortal souls. Those who remain at home have not that happiness....Let us then hear no more of this detestable proposition [the freeing of the slaves], the adoption of which would, by depreciating our lands and houses and thereby depriving so many good citizens of their properties, create universal discontent."

Eliza Powel became both a social and political adviser to General Washington.

Washington described the garden "of lemon, orange, and citron trees, and many aloes and other exotics."

Robert Morris was said to be the richest man in the city. Morris certainly lived in the most sumptuous house. It was an easy walk from Morris's house to Ben Franklin's, or the State House, or the Powels, or the Indian Queen. Morris, you remember, was the man who had the hard job of trying to pay the Revolutionary war bills.

His house was full of history. William Penn's son had built it; General Howe had lived there during the recent war; now George Washington was staying in a second-floor room. Morris was pleased to have the most famous man in America as a guest.

James Wilson's house was diagonally across Market Street from the Morris house. Wilson, another Pennsylvania delegate, was a bookish man, like Madison. He had come to America from Scotland during the Stamp Act crisis, and was soon a national leader. When Governor John Rutledge's ship docked in Philadelphia, a message from James Wilson was waiting for him. Wilson invited Rutledge to stay in his house and, until Mrs. Rutledge arrived three weeks later, he did.

Their friendship was surprising; perhaps each admired the lawyerly skills and no-nonsense intelligence of the other. "Interest alone is the governing principle of nations," said Rutledge. By interest he meant property, and property included slaves.

Politically, James Wilson stood at the other end of the table. People and their individual rights were more important to him than property rights. He opposed slavery and favored democracy. Look at a portrait of Wilson and you will see round glasses, powdered hair, and plump cheeks. His clothes are dark and plain—nothing fancy about that man. His mind, however, was described as a "blaze of light."

James Wilson was a Scot, a successful lawyer, and a man who spoke his mind.

Wilson was for power for the people. Not everyone believed in democracy. That word had a different ring in the 18th century from what it would have in later times. Many felt the people could not be trusted; they feared mob action and thought only the wealthy and the educated should vote. There was something else that made Wilson's thinking different from that of most of the delegates. He cared little for the states. It was the new nation that he thought important. "When Wilson speaks," a historian has written, "he wastes no time and considers no man's feelings." Not everyone liked James Wilson; all respected him.

Gouverneur Morris lost his leg in a coach accident. Driving was hazardous in the 18th century too.

Now Gouverneur Morris, who was no kin to Robert Morris, but from the same state, was well liked. He was bold, witty, dashing, and also brilliant. Morris had a reputation for loving the ladies, and they seem to have returned the feeling. He was as tall and well made as George Washington, and a bit vain about it. He let people know that a sculptor had asked him to model for a statue of Washington. That Morris had a wooden stick in place of one leg didn't seem to make any difference to the ladies or to the sculptor.

The Morrises—Gouverneur (left) and Robert (right)—were from the same state, Pennsylvania, but not related.

Morris was more than just good-looking. He spoke often during the Convention—more than anyone else—and he spoke well. When the Constitution was finally almost finished—it had been argued and re-argued and changed and re-changed—it was to Morris that they gave it. He was to polish the words. He could write with skill and grace, and all knew it. He took 23 resolutions and reduced them to seven articles. The language he used was forceful, clear, and tight. They picked the right man.

Did you guess already? It was Gouverneur Morris who slapped

Alexander Hamilton as a dashing young staff officer in the Revolutionary War. He never got his degree from King's College, because he dropped out to fight.

George Washington on the back. Alexander Hamilton dared him to do it. Hamilton bet Morris a dinner that he wouldn't go up to the reserved and formal Washington, throw his arm around him, and say, "How are you today, my dear General?" (Washington, by the way, was 20 years older than Morris.)

Morris did it, but afterward he said—when he saw the startled look on Washington's face—that he wouldn't do it again for a thousand dinners.

Alexander Hamilton knew George Washington well. He had been a member of Washington's staff during the Revolution. Hamilton was 35, and the youngest delegate. Like Gouverneur Morris, he wrote well and was dashing and handsome. Like James Madison, he was small and slim. Madison and Morris came from privileged homes; Alexander Hamilton was a poor boy from a broken home.

His father had deserted the family; his mother died when he was 11. But Hamilton was a genius, and at 13 he was helping manage a store and using two languages to do it. He grew up in the Virgin Islands; a hurricane blew him north. Hamilton wrote an article about the hurricane that was published in a newspaper. The article was so well written that people raised money to send Hamilton to King's College in New York (now known as Columbia University). To enter King's College you had to pass an exam in Latin and Greek. Hamilton passed.

John Dickinson changed his mind about the states later. He said, "We must either subject the states to the danger of being injured by the power of the national government or the latter to the danger of being injured by the states. I think the danger is greater from the states."

He came to the Constitutional Convention as a delegate from New York. His ideas were different from those of many of the delegates. Hamilton wanted the United States to have a government like England's. He thought it the best possible model. He wanted the president to have a lifetime job—like a king or an emperor. Hamilton wanted a strong central government. He didn't care about the states' having power.

The delegates were divided on many issues—but most of all on *power*: who should have it and how much? Some delegates wanted the states to be strong; others were for a strong national government; still others hoped for a balance.

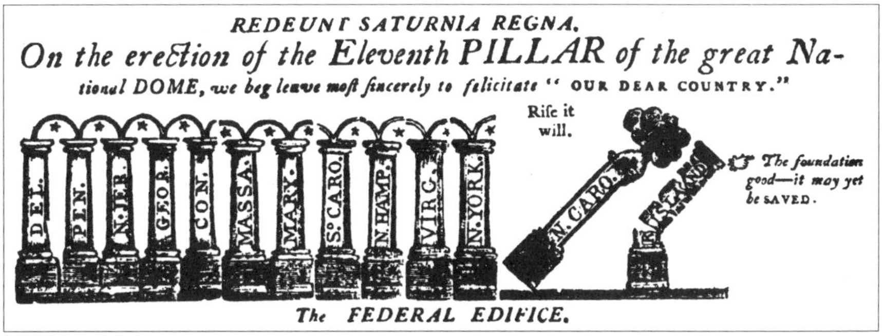

It took the states even longer to agree to the Constitution than it did the delegates. This cartoon shows wobbly North Carolina and Rhode Island, the last states to ratify.

John Dickinson, of Delaware, was for strong states. Do you remember him? Dickinson was the man who wrote the Articles of Confederation. And who refused to sign the Declaration of Independence. Dickinson had powerful opinions. He wanted a *confederation*. Madison wanted a *federation*. Those are confusing words, so let's define them.

A *confederation* is a government made up of a group of partners. Those partners keep all important power for themselves. That's what happened under the Articles of Confederation. There was no higher power. The central government served mainly as an adviser.

A *federation* is a form of government that divides power between a central government and state governments. It isn't easy to do that. The United States is a federation. It has a *federal* form of government. The central government in Washington has the strongest powers, but not all the power. The states have important powers, too. It is a balanced form of government.

Our government in Washington isn't as strong as Hamilton wanted. The states' powers aren't as strong as Dickinson wished. The Constitution was a compromise. Both sides gave and both sides got. It may have been the best compromise in history. It certainly is the best constitution.

Pounds are a unit of British money, as dollars are of American money. ***Inculcate*** means to teach. A ***commencement*** is a beginning. ***Parochial*** can mean provincial or limited; here Franklin simply means belonging to the parish, the local community.

Sense Being Preferable to Sound

In 1785 Ben Franklin was asked to donate a bell to a Massachusetts town named in his honor. He was in France when he wrote in reply to the request:

My nephew, Mr. Williams, will have the honor of delivering you this line. It is to request from you a list of a few good books to the value of about twenty-five pounds, such as are most proper to inculcate principles of sound religion and just government. A new town in the State of Massachusetts having done me the honor of naming itself after me, and proposing to build a steeple to their meeting house if I would give them a bell, I have advised the sparing themselves the expense of a steeple at present, and that they would accept of books instead of a bell, sense being preferable to sound. These are therefore intended as the commencement of a little parochial library for the use of a society of intelligent, respectable farmers such as our country people generally consist of... With the highest esteem and respect, I am ever, my dear friend, yours most affectionately,

B. Franklin

37 Roger to the Rescue

William Paterson, said another delegate, was "a classic, a lawyer, and an orator."

Paterson, New Jersey, is named after William Paterson.

There was one issue on which everyone was stubborn, and it had to do with the legislative branch. The fight was between the big states and the little ones. No one would give in.

The Virginia plan said that the number of congressmen each state would have should be decided by population. There is some sense in that—but, of course, it favored the states with the most people: Virginia, Massachusetts, and Pennsylvania. These states would have most of the congressmen.

The New Jersey plan was introduced by William Paterson. It said each state should have an equal number of representatives in Congress. That meant that Delaware, with 59,000 people, would have the same number of congressmen as Virginia, with almost 692,000. Was that fair?

Neither side would budge. It seemed that a constitution would never get written. Some people talked of quitting and calling another convention to try again. Then Roger Sherman came up with a compromise.

Sherman was a tough old Connecticut Yankee: a lean, sharp-nosed man with big hands and feet who wore plain, sensible clothes and spoke only plain, sensible words. Georgia's delegate described Roger Sherman as "the oddest shaped character I ever remember to have met with. He is awkward...and unaccountably strange in his manner ...[yet] no man has a better heart nor a clearer head."

Roger Sherman had signed the Declaration of Independence and the Articles of Confederation. At age 66, he was the second oldest man at the convention. As a young man he had been a shoemaker and a

farmer; then he taught himself law and became a lawyer. John Adams called him "an old Puritan, honest as an angel."

Thomas Jefferson once pointed to him and said, "There is Mr. Sherman of Connecticut, who never said a foolish thing in his life."

In Connecticut, so the story goes, he was asked to give the dedication speech for a new bridge. He walked out onto the bridge, stood for a moment, came back, and turned to the crowd. "It stands steady," he said. That was the whole speech.

The Constitutional Convention needed a man of good sense and few words. Here is Roger Sherman's compromise (actually, it is known as the Connecticut Compromise):

One house of the legislature should reflect a state's population—the House of Representatives.

One house should have an equal number of representatives from each state—the Senate.

That was it. That simple solution meant there would be a Constitution. After that, it was just a matter of details.

Roger Sherman once said, "The people should have as little to do as may be about the government. They...are liable to be misled."

38 Just What Is a Constitution?

Three-Fifths of a Person

When it came to counting people for representation in Congress, the states with large slave populations wanted to count their slaves just as they did their free citizens. That would give them a large number of representatives. The people in the states with few slaves objected. They said, "Slaves don't have rights as citizens. They shouldn't be counted at all." There was much argument about this and finally a compromise. Here it is: Article I, Section 2 of the Constitution says: *Representatives...shall be determined by adding to the whole Number of free Persons...three fifths of all other Persons.* In other words, each slave, for counting purposes, was to be tabulated as three fifths of a person. Does that sound horrible? Well, those who hated slavery saw it as a partial victory. They had not let the South Carolina and Georgia extremists have their way.

This eagle with 13 stars, an olive branch, a sheaf of arrows, and the sign *E Pluribus Unum* ("out of many, one"), became the United States' official seal in June 1782.

We've been talking about constitutions, but do you actually know what a constitution is? Or what legislation is? Or how they differ from each other?

Well, legislation means laws. And laws change, sometimes with each generation. Traffic laws that worked in horse-and-wagon days aren't right for times with fast automobiles and airplanes. Our Constitution, however, is made up of superlaws, which are *not* meant to be easily changed. The Constitution has gone from the time of candlelight into the age of rockets with only a few changes. How come?

Because a constitution—a good constitution—is just a basic plan that helps people live together in peace and happiness. It provides a way for people to make everyday laws and enforce them. Laws change with the times; a good constitution shouldn't need much changing.

Why don't you make a constitution for your classroom or your home? Remember, you have adults and children who need to agree to this constitution. Should each have an equal vote? Can you have a perfect democracy, or isn't that suitable? How will you balance power? Who gets to make the laws? Everyone? One person? The teacher? A chosen few? What happens to anyone who breaks a law? Does a visitor have to obey the rules? Suppose you want to change your constitution? How would you do that? Will your constitution be respected if it can be easily changed?

The Constitution writers asked those same kinds of questions. But, as you can imagine, when you write a constitution for a big country you have a lot of organizing to do—especially if you want to be fair. And more than anything else, the men who wrote our constitution wanted to be fair.

One of the nice things about our Constitution is that it is not terribly complicated. You won't find it as exciting as *Tom Sawyer*, but if you read it carefully, you will understand it.

It is that simplicity that has helped make the Constitution so lasting. The delegates to the Convention came up with a constitution that is still great more than 200 years after it was written. No other country has ever had a governing document that has worked so well for such a long time.

One of the first things the delegates decided on was a name for the new country. The Convention officially adopted the name the *United States of America*.

This 1789 print celebrates George Washington and the 13 states. Each coat of arms gives the population of its state, and the four corners of the picture contain information about the political structure of the new country.

Then the delegates agreed to a three-part government with legislative, executive, and judicial branches. That was part of Madison's Virginia Plan. It was based on the English plan of government.

A legislature is a lawmaking body: a congress or parliament. Our Congress is divided into two groups, called houses. They are the Senate and the House of Representatives.

An executive is a leader: a president or king.

The judiciary is the courts. In the United States they go from town courts all the way to the Supreme Court.

Since the delegates were afraid of power, the three branches were planned to check and balance each other. They actually used that phrase: *check and balance*. Think about what it means. They expected

The lawmaking body in Germany is the Bundestag. In Japan it is the Diet. In Israel it is the Knesset. What is England's congress called? Do you know any others?

each branch to keep the others in check—to stop them going too far and overstepping the limits of their authority. And balance? Well, no branch was supposed to be stronger than any other.

The president was made commander in chief of the army and navy, but was not given the power to declare war. Only Congress can do that. That was one way to balance power.

Power was balanced between the states and the national government. The national government controls foreign affairs, business between the states, and the post office. The states control schools, roads, and local government.

Congress was given the power to impose taxes. Control of money is an important power.

Organizing the details of the three branches took a lot of time. The

The Pennſylvania Packet, *and Daily Advertiſer.*

[Price Four-Pence.] WEDNESDAY, September 19, 1787. [No. 2690.]

WE, the People of the United States, in order to form a more perfect Union, eſtabliſh Juſticc, inſure domeſtic Tranquility, provide for the common Defence, promote the General Welfare, and ſecure the Bleſſings of Liberty to Ourſelves and our Poſterity, do ordain and eſtabliſh this Conſtitution for the United States of America.

ARTICLE I.

delegates argued over everything. Hamilton wanted the president to be like a king. Edmund Randolph wanted a committee of three to act as president. Ben Franklin wanted a legislature with only one house. They voted 60 times before they could agree on the way to select a president.

Would the legislators and the judges be elected or appointed? How long would they serve? What would they be called? Those questions needed answers.

Someone suggested the president be called Your Mightiness. Do you know what the President is called?

Our Constitution was the work of sensible men who didn't think our leaders needed fancy titles. They were students of government:

If you listen to a press conference you will hear his official title. It is *Mr. President*. When a woman is elected president, what do you think she will be called?

almost all of them had done a lot of reading and studying. Most had served in their state legislatures. They had helped write their state constitutions. They disagreed on many things, but on two ideas all were agreed:

•They wanted to *guarantee basic human rights and freedom* (what Jefferson called "unalienable rights").

•They wanted to *provide government by consent of the governed*. That means they expected the people to govern themselves through their representatives.

Those don't seem like unusual goals to us today, but they were unusual in the 18th century. No nation had ever done what those men hoped to do. England had come closest. But the men who were writing our constitution—the Framers—knew all about the problems in England in times when the king had too much power and in other times when Parliament had too much power.

Finally, the delegates solved the problem of power in two ways: with checks and balances (as I just told you), and by making the Constitution more powerful than any president, congress, court, or state.

They made the Constitution the supreme law of the land. None of the three branches is allowed to break its rules.

But the Constitution isn't a rigid, unbending document. Those very wise Framers made it flexible. They came up with a way of changing the Constitution in order to adapt it to new times and new ideas.

The Constitution can be changed by *amendments*. However, the Framers made it hard to do that. They didn't want people to change the Constitution just to keep up with passing fads. Amendments must be approved by Congress and by the people in their state legislatures. Two-thirds of the members of Congress need to vote for an amendment for it to be approved, and three-quarters of the state legislatures must ratify it. So far, 10,000 amendments have been suggested for our Constitution—only 26 have been approved.

Good as it was, the new Constitution wasn't perfect. The delegates made a few bad mistakes. Thank goodness for that amendment process. It would help to correct them.

Please, Say It Like It Is!

Here's something you may notice when you read the Constitution. The words *slave* and *slavery* are never used. The Constitution writers substituted "person held to service or labor," or "all other persons," for *slave*. When you substitute a pleasant word for an unpleasant or evil one, you are using a *euphemism* (YOO-fuh-miz-um). The word *euphemism* comes from Greek roots: *eu* meaning good, and *pheme* meaning words or speech. We use euphemisms all the time: a toilet is a bathroom, a rest room, or a powder room—all "good" words. We avoid the unpleasant subject of death by saying that someone "passed away." People who clean sewers are not called sewer cleaners; they are sanitary engineers. Can you think of other common euphemisms?

39 Good Words and Bad

John Rutledge had been a young man back in 1765, when he met William Johnson. Now he was 48 and a governor.

We the People of the United States, in order to form a more perfect union, establish justice, insure domestic tranquillity, provide for the common defence, promote the general welfare, and secure the blessings of liberty to ourselves and our posterity, do ordain and establish this Constitution for the United States of America.

Those were Gouverneur Morris's words, and they were just right. Morris sometimes talked for hours, but when he wrote the Constitution he wrote just enough—and no more. That opening paragraph, called the *preamble* (PREE-am-bull), is worth memorizing. Don't you agree, it is fine writing?

But when the Framers said "we the people," did they really mean all the people? I'm sorry to have to tell you that most experts say no.

They say that the Founders didn't mean women, who were not allowed to vote. And they didn't mean Native Americans or blacks.

"I came here," said Gouverneur Morris, "as a representative of America. I... came here in some degree as a representative of the whole human race."

The Constitution was different from the Declaration of Independence. The Declaration stated goals, but the Constitution was concerned with what would actually be done by the government.

A slave auction in progress. "We have," said James Madison at the Convention, "seen the mere distinction of color made, in the most enlightened period of time, a ground of the most oppressive dominion ever exercised by man over man."

Now, hear this, because there is more than one side to this question, and you will need to form your own opinion. I don't agree with the experts. I think that when the Founders said "we the people," they meant *all* the people.

Those Founders—most of them, anyway—were idealistic men. They were thinkers who were ahead of their times. They knew they were writing for future generations. So they wrote the greatest document they could write.

But they were also practical men. They had to create a working document that people would approve. And they knew that some of the citizens of the new United States—and even some of the delegates to the Constitutional Convention—weren't prepared to do some things that had never been done before. They weren't prepared to accept women as citizens. They weren't prepared to give up property when that property was a slave. They weren't prepared to be fair to Indians in the rivalry over land. So, many of the delegates acted as if "we the people" meant "we the grown-up white men who own property." They were compromising. Maybe they had to compromise. Maybe they thought the problems could be solved later.

Here is something ironic. Native Americans usually began their treaties with that phrase, "we the people." Did the men at the Constitutional Convention know that?

South Carolina's John Rutledge had been a good friend of Sir William Warraghiyagey Johnson (who, as you know, was now dead). It is said that Rutledge brought a copy of an Iroquois treaty to the Constitutional Convention. But no one knows that for sure. Remember, George Washington insisted that the sessions be secret. Wherever they came from, the words were perfect. This was to become a people's government, even though, in the beginning, some people were left out. But it was those three small words, "we the people," that kept pushing the nation to include all peoples.

Now consider this: in a democracy, politicians usually do only what the people want. The settlers wanted Indian lands. They would have them.

Could a way have been found to share the land and please everyone? Perhaps not. But that doesn't mean a better way couldn't have been found. The Framers—and most of the people they represented—didn't know how to live with the Indians as partners in a new nation. They didn't even offer them citizenship. (Later, the 14th Amendment to the Constitution granted citizenship to everyone born in this country. But it was not until

Washington owned slaves himself. Behind him, a slave holds the general's horse. Like many, he was willing to allow slavery in order to get the Constitution approved by all.

1968 that Native Americans living under tribal law on reservations were guaranteed the full rights of American citizens.) The Framers did not come up with a fair way to share the land with the Native Americans.

The ideas of the Enlightenment—the 18th-century belief in the rightness of human reason—made people think hard about slavery. Fifty years earlier, hardly anyone had worried about whether slavery was wrong. By 1787 almost everybody knew it was.

Another big mistake was made. The Constitution didn't outlaw slavery.

Most of the delegates didn't want slavery. They knew it was wrong. But in the South—especially in South Carolina and Georgia—a way of life depended on slave labor. The citizens of those states would not approve the Constitution if it prohibited slavery. They threatened to stay independent. Some of the other slaveholding states might follow them. Then the North American continent might end up like Europe—with many nations instead of united states. Some of those nations would be free nations, and some would allow slavery.

If you were a delegate, what would you do?

The delegates gave in. They did what the Deep South wanted. They agreed to let slavery continue. They even allowed the dreadful slave trade to continue.

The slave trade was the business of bringing Africans into the country and selling them as slaves. Virginia's George Mason knew the South was not ready to eliminate slavery. He thought, as a first step, that the slave trade should be outlawed. He wanted no more people to be enslaved. The delegates compromised on that. The Constitution gave the slave traders another 20 years—until 1808—and then that diabolical and disgraceful business was to be stopped.

"Slavery," said George Mason, "discourages arts and manufacturing...[and] every master of slaves is born a petty tyrant."

Sometimes it is hard to see things with the eyes of another age. But if you want to understand history, you have to do that. Try it right now. Imagine that you are in Philadelphia in 1787, listening to James Wilson, Pennsylvania's scholarly delegate. Wilson is proud that Pennsylvania no longer allows slavery. He is talking about the Constitution:

Diabolical means devilish or fiendish.

> *It gives me great pleasure that so much was done....I consider this as laying the foundation for banishing slavery out of this country; and though the period is more distant than I could wish, yet it will produce the same kind of gradual change which was pursued in Pennsylvania.*

It turned out that Wilson was wrong. Slavery would not disappear gradually. But, in 1787, many wise people thought slavery was dying. Slavery seemed to be becoming unprofitable. Most white people thought slavery would just disappear naturally. They didn't think it worth a fight. But they were wrong. They couldn't know there would

Elbridge Gerry didn't want to have a vice president. "The close intimacy that must subsist between the President and Vice President makes it absolutely improper," he said. But he became James Madison's vice president!

Paradoxical means puzzling. A ***paradox*** is something that is the opposite of what you might logically imagine it to be.

soon be an invention—the cotton gin—that would change farming in the South. It would make slavery profitable again.

So they did what was practical, not what was right. Elbridge Gerry, a prosperous merchant from Marblehead, Massachusetts, saw the price the nation would eventually pay for their practicality. In a letter to his wife he wrote, "I am exceedingly distressed at the proceedings of the Convention—being...almost sure, they will...lay the foundation of a Civil War." Gerry had signed the Declaration of Independence and the Articles of Confederation. You'll soon see what he did about the Constitution.

Understanding history is hard because it is often paradoxical. So are people and nations. It would be easier but less interesting if people were either all good or all bad, and if the good people never made mistakes.

And, talking of a paradox, you can't find a better example than South Carolina's John Rutledge. Rutledge wouldn't budge an inch when it came to defending South Carolina's interests. That means he fought to keep slavery. He fought harder than anyone in Philadelphia. But while he was arguing in support of slavery at the convention, he was quietly freeing his own slaves. Rutledge had inherited 60 slaves; he owned only one when he died. His wife freed all her slaves. George Mason and Thomas Jefferson hated slavery and said so. But they owned slaves and kept them. Now that is a bit of a paradox, isn't it?

A Taste of Freedom

It was August 1793, and some of Philadelphia's leading white citizens were sitting at a long plank table under some shady trees. They were feasting on meats and pies and juicy melons. Black men served the bountiful meal.

When they were finished, the white men stood up and the black men sat at the banquet table and were served a similar meal by "the most respectable of the whites." What was this all about?

These men had seen the roof beams raised on the first free black church in the new American republic, the African Church of Philadelphia; now they were celebrating together. "Peace on earth and good will to men," said Benjamin Rush. One of the black leaders, William Gray, tried to thank the whites for their help, but he "was checked by a flood of tears."

Afterwards, Rush told his wife, there was much shaking of hands and "virtuous and philanthropic joy." The men—black and white—believed they could live in harmony. They were tugging at that ladder of society. They expected to turn it on its side. (This was a beginning; perhaps at the next dinner they would all eat together.) They thought that slavery and racial separation would soon be things of the past. It turned out that they were wrong.

"Not until the latter part of the twentieth century would Philadelphians and the generality of other Americans work their way back to where they had stood two centuries before—perceiving the possibility of a society undivided by race," said historian Gary Nash in a book called *Forging Freedom.*

40 No More Secrets

When John Hancock and Samuel Adams were reluctant to support [the Constitution's] ratification, Paul Revere organized the Boston mechanics into a powerful political force, and worked behind the scenes with such effect that he is commonly thought to have turned the narrow balance in a critical state. Once again, he showed his genius for being at the center of great events.

—David Hackett Fischer, *Paul Revere's Ride*

The imposing chair, decorated with a half-sun, that Washington sat in during the Convention.

What happens when you have a secret? Do all your friends want to know about it? Of course they do. People are always curious when there is a secret.

All that summer of 1787, everyone was curious about what was going on at the Convention. Even James Madison's father complained. He suggested that his son should at least tell him what had *not* been decided.

But George Washington meant it when he said, "No talking." One delegate dropped his notes on the floor, and Washington lectured everyone on carelessness.

A two-headed snake was found in a river and sent to the scientific-minded Dr. Benjamin Franklin. Franklin showed it to a friend and forgetfully compared it to a convention being pulled in two directions. He was stopped in mid-sentence by another delegate who feared he would say too much.

At the State House they even kept the windows closed so no one could stand outside and listen. Some delegates almost fainted from the

At the Convention's end, Benjamin Franklin looked at the chair's ornament and said, "I have the happiness to know that it is a rising and not a setting sun."

heat before they all moved upstairs, where they could open windows and not be heard on the street.

Finally, the need for secrecy was over. On September 17, 1787, the Constitution was finished and ready to be signed. Two Virginians—Edmund Randolph and George Mason—and a Massachusetts man—Elbridge Gerry—left without signing. They didn't think it good enough.

It wasn't perfect, said Ben Franklin, but it was better than he had expected. "It astonishes me," he said, "to find this system approaching so near perfection as it does; and I think it will astonish our enemies." Tears are said to have streamed down Franklin's cheeks as he signed his name. He pointed to the chair where George Washington had sat all summer as president of the Convention. Carved on the chair's back was a half-sun with sunbeams. Franklin said he had often wondered if it were a rising or a setting sun. Now he knew: the sun was rising.

Washington wrote to Lafayette. "It appears to me, then, little short of a miracle," he said in his letter, "that the delegates from so many states...should unite in forming a system of national government."

When Patrick Henry was asked why he stayed home, instead of helping to make a good constitution, he said, "I smelled a Rat."

At last the newspapers could print the Constitution and write about it; everyone could read it and form an opinion. All the arguing that the delegates had done was now repeated by citizens all over the land.

That old firebrand Patrick Henry hated it. Before the Convention even began, Henry said he "smelled a rat in Philadelphia." The rat was the whole new Constitution. Henry thought the Articles of Confederation should have been revised. He didn't want a federal republic; he wanted powerful states.

Those words "we the people" enraged Henry. "It should be *we the states!*" he bellowed.

Each state called a special convention to decide whether to ratify—that means approve—the Constitution. The debates were hot at those conventions.

In Virginia, Patrick Henry was up to his old speechmaking tricks. He was still a spellbinder. Picture this: Patrick Henry so excited trying to defeat the Constitution that he twirled his wig around on his head—more than once. He almost won. But small, soft-voiced James Madison spoke up. Their liberties would be safe under the Constitution, he told the citizen-delegates. They knew that George Washington approved of the document. That convinced them, even though the respected George Mason didn't agree.

Two weeks after Maryland signed the new Constitution, a writer in the *Maryland Journal* had some second thoughts. Why didn't the Constitution outlaw slavery, he wondered? "The members of the...convention should have seized the happy opportunity of prohibiting forever this cruel species of reprobated villainy," he wrote.

I told you that George Mason was one of the three delegates who refused to sign the Constitution. Mason had good reasons. He hated slavery and wanted at least to end the slave trade. He wouldn't compromise.

But most of the delegates, like Benjamin Franklin, remembered the old proverb: half a loaf is better than no bread at all. They knew the Constitution was an amazing document. It had faults, but it also had the cure for those faults in a built-in amendment process. The American people, with much wisdom, would make use of that amendment process. In time they would fix the faults. They began at once.

Washington and the delegates debating the Constitution and its compromises.

41 If You Can Keep It

James Madison wrote the Bill of Rights and got Congress to pass it.

George Mason was right about something else. "The Constitution has no bill of rights," he said.

Now how could that have happened? All those brilliant men and they forgot a bill of rights?

Well, they hadn't forgotten. They just thought that because the state constitutions had bills of rights, a general one wasn't necessary. But it was. A good bill of rights guarantees things like freedom of religion, freedom of the press, and free speech. Americans cared deeply about those freedoms.

Mason wasn't the only one who was upset. Jefferson wrote from Paris to Madison and Washington that a bill of rights was needed. In Massachusetts and Virginia the state conventions agreed to ratify the Constitution only after they added strong recommendations that the first Congress pass a bill of rights. North Carolina and Rhode Island wouldn't even ratify the Constitution. They were afraid of losing precious liberties if they joined the infant nation. They wanted to see a bill of rights before they signed.

This is what happened: James Madison served in the first Congress and he wrote the first 10 amendments to the Constitution. They are called the Bill of Rights.

The First Amendment guarantees freedom of religion, freedom of speech, freedom of the press, and freedom to assemble (to gather in groups) and protest or petition the government when there are problems to be solved. Most people agree: the First Amendment is the most important part of the Bill of Rights.

The Second Amendment says, "A well regulated militia, being necessary to the security of a free state, the right of the people to keep

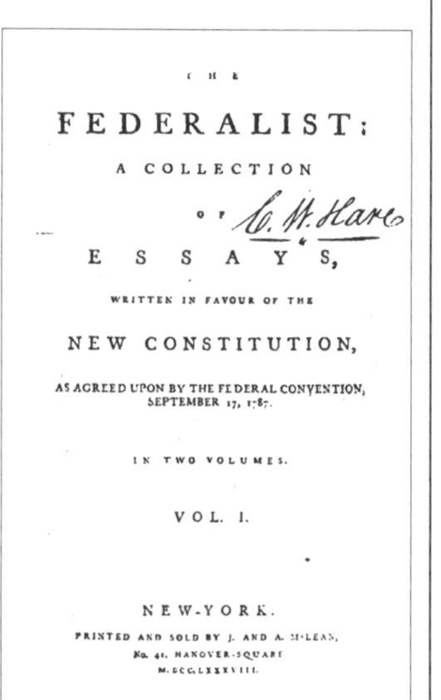
THE

FEDERALIST:

A COLLECTION

OF

ESSAYS,

WRITTEN IN FAVOUR OF THE

NEW CONSTITUTION,

AS AGREED UPON BY THE FEDERAL CONVENTION, SEPTEMBER 17, 1787.

IN TWO VOLUMES.

VOL. I.

NEW-YORK:

PRINTED AND SOLD BY J. AND A. M'LEAN,

No. 41, HANOVER-SQUARE

M.DCC.LXXXVIII.

Hamilton, Madison, and John Jay wrote a series of 85 articles to convince New York's voters to approve the new constitution. They are called the *Federalist Papers*.

In 1789 an unknown artist celebrated the new nation with this symbolic picture of *America Trampling on Oppression.*

and bear arms, shall not be infringed." To "bear arms" means to carry guns. But there is more to it. Read the whole amendment. Do we have a militia today? Why was it important for citizens to bear arms in the 18th century? Do you think it is important today?

The Third Amendment says that soldiers can't be quartered in your house without your consent.

The Fourth Amendment says that the police can't come and search your house unless they have proper legal papers. (In some countries the police can burst in on you at any time. Would you like that?)

The Fifth, Sixth, Seventh, and Eighth Amendments all have to do with fair trials. The Fifth Amendment says no person can be forced to be a "witness against himself." That helps protect us from those who might be tempted to use torture to force confessions.

The Ninth and Tenth Amendments say that any powers or rights that the Constitution does not give to the federal government belong to the states or to the people. Why do you think the Framers wrote those amendments? (Recent court decisions have said that the Ninth Amendment gives people the right to privacy. That means we can have secrets—even from the government.)

Read the Bill of Rights at the end of this chapter. It gives you freedoms that people in many nations would love to have. You should know about them.

Can you guess what Patrick Henry did when Virginia's General Assembly voted approval of the Constitution? Well, he didn't sulk. He said if it was the will of the people then he was for it.

Alexander Hamilton didn't get the kind of constitution he wanted. He didn't get a king-like president. But when the Constitution was signed, Hamilton supported it. He even wrote a series of articles that were published in newspapers. They helped convince people to vote for the Constitution. Henry and Hamilton were gracious in their actions, and that became the American way.

In this country, when a candidate loses an election for president, he doesn't become spiteful or nasty. He pledges to help the new president. That is the American way.

Washington was elected president unanimously. He longed for the peace of his estate in Virginia, but he agreed, once again, to serve his country. That, and the Bill of Rights, convinced the people of North

A ***republic*** is a government in which power is held by citizens who vote and elect their representatives to make laws and govern the country.

Carolina. North Carolina ratified the Constitution and became part of the United States. Independent-spirited Rhode Island was the only state that never sent a delegate to the Constitutional Convention. But, finally, the Rhode Island legislature voted—34 to 32—to ratify the Constitution. Now there were 13 states, and just about everyone was feeling good about the new nation and confident with George Washington as first president.

Except people in Europe. They said it wouldn't last. They predicted that the country would soon fall apart. A king or dictator would be necessary. It was all a wild experiment, they said. Self-government? People can't be trusted to govern themselves. People writing their own constitution? What an impractical dream!

We fooled them. We showed it could work. The United States proved the worth of self-government. We proved the worth of democracy.

Will our nation continue to last free and democratic? Jefferson knew that self-rule depends on informed citizens. He believed in education. He believed in freedom of the press. Will your generation be informed? Will you preserve American democracy?

The birth of the United States gave rise to many pictures like *Liberty as Goddess of Youth* (1796)—where youthful liberty feeds the American eagle.

When Benjamin Franklin came out of the Pennsylvania State House, on September 17, 1787, his friend Elizabeth Powel, the wife of the mayor of Philadelphia, was waiting for him. She asked what kind of government the new nation would have.

"A republic, madam," he told her. "If you can keep it." Those words were not meant just for Mrs. Powel. They were also meant for you.

WHEW – It was done.
The CONSTITUTION written,
the U.S. begun.
The world wasn't upside down at
all. No way!
A NEW AGE had come to stay.
So read on, good friends,
in Book Four,
The NEW NATION will grow,
and fight, and more.

The Bill of Rights

Amendment I

Congress shall make no law respecting an establishment of religion, or prohibiting the free exercise thereof; or abridging the freedom of speech, or of the press; or the right of the people peaceably to assemble, and to petition the Government for a redress of grievances.

Amendment II

A well regulated Militia, being necessary to the security of a free State, the right of the people to keep and bear Arms, shall not be infringed.

Amendment III

No Soldier shall, in time of peace be quartered in any house, without the consent of the Owner, nor in time of war, but in a manner to be prescribed by law.

Amendment IV

The right of the people to be secure in their persons, houses, papers, and effects, against unreasonable searches and seizures, shall not be violated, and no Warrants shall issue, but upon probable cause, supported by Oath or affirmation, and particularly describing the place to be searched, and the persons or things to be seized.

Amendment V

No person shall be held to answer for a capital, or otherwise infamous crime, unless on a presentment or indictment of a Grand Jury, except in cases arising in the land or naval forces, or in the Militia, when in actual service in time of War or public danger; nor shall any person be subject for the same offence to be twice put in jeopardy of life or limb; nor shall be compelled in any criminal case to be a witness against himself, nor be deprived of life, liberty, or property, without due process of law; nor shall private property be taken for public use, without just compensation.

Amendment VI

In all criminal prosecutions, the accused shall enjoy the right to a speedy and public trial, by an impartial jury of the State and district wherein the crime shall have been committed, which district shall have been previously ascertained by law, and to be informed of the nature and cause of the accusation; to be confronted with the witnesses against him; to have compulsory process for obtaining witnesses in his favor, and to have the Assistance of Counsel for his defence.

Amendment VII

In suits at common law, where the value in controversy shall exceed twenty dollars, the right of trial by jury shall be reserved, and no fact tried by a jury, shall be otherwise reexamined in any Court of the United States, than according to the rules of the common law.

Amendment VIII

Excessive bail shall not be required, nor excessive fines imposed, nor cruel and unusual punishments inflicted.

Amendment IX

The enumeration in the Constitution, of certain rights, shall not be construed to deny or disparage others retained by the people.

Amendment X

The powers not delegated to the United States by the Constitution, nor prohibited by it to the States, are reserved to the States respectively, or to the people.

Chronology of Events

1215: the English barons force King John to sign Magna Carta
1688: England's Glorious Revolution makes Parliament more powerful than the king
1735: John Peter Zenger is acquitted of libel for printing criticism of New York's British governor
1738: the Spanish cut off Captain Jenkins' ear
1739: the Great Awakening, a religious revival, begins
1754: Col. George Washington and 150 Virginians defeat a French exploratory party in Pennsylvania and start the French and Indian War
1754: the colonies reject the Albany Plan, Benjamin Franklin's proposal for a union
1755: General Braddock's British troops are defeated by French and Indians at Fort Duquesne
1755: William Johnson and Hendrick defeat the French at Lake George
1755: the Acadians leave Nova Scotia because they will not swear loyalty to Britain
1759: the British are victorious at Quebec
1760: Johnson and General Amherst capture Montreal. The war in America is won by the British
1760: George III becomes king of England
1763: the Treaty of Paris ends the Seven Years' War
1763: the Proclamation of 1763 forbids American colonists to settle west of the Appalachians
1765: the Stamp Act forces the colonies to pay taxes on printed matter
1765: the Sons of Liberty band together against taxation without representation
1766: Parliament repeals the Stamp Act
1767: Parliament passes the Townshend Acts, which tax tea and other goods
1769–82: Father Junipero Serra and other Franciscan friars establish 21 missions in California
1770: five Americans killed in the Boston Massacre
1773: the Boston Tea Party: 50 rebels throw chests of tea into Boston harbor to protest the tax
1773–74: Britain blockades Boston harbor
1774: the First Continental Congress meets in Philadelphia to protest and petition George III
1775: the battles of Lexington and Concord
1775: the Second Continental Congress meets in Philadelphia and names George Washington commander in chief of the American forces
1775: Patriots under Ethan Allen and Benedict Arnold take Fort Ticonderoga
1775: Britain declares war on America
1775: the battle of Breed's and Bunker hills in Boston
1776: Tom Paine publishes *Common Sense*
1776: the Declaration of Independence
1776: the British capture New York City. The Americans retreat to Pennsylvania
1776: American troops defeat the British at Fort Sullivan in Charleston, South Carolina
1776–77: Washington crosses the frozen Delaware and captures Trenton and Princeton, New Jersey
1777: the Americans under General Gates defeat General Burgoyne at Saratoga
1777–78: the Americans winter at Valley Forge
1778: George Rogers Clark captures forts at Cahokia, Kaskaskia, and Vincennes in the Ohio Valley
1781: the French fleet and American troops defeat the British at Yorktown. The war is over
1781: Congress adopts its first constitution, the Articles of Confederation
1783: Britain recognizes American independence
1787: the Northwest Ordinance divides up the Northwest Territory among several states. It forbids slavery in these areas
1787: the Constitutional Convention adopts a new Constitution
1788: the Constitution is ratified by three-quarters of the states and becomes law
1789: George Washington is elected first president of the United States
1791: James Madison writes the first 10 amendments to the Constitution, the Bill of Rights

More Books to Read

At the end of a history book, most writers make a list of books for further reading. I decided I would tell you about some books that I *love* to read.

James Lincoln Collier & Christopher Collier, *My Brother Sam Is Dead,* Four Winds Press, 1974. Tim Meeker is a boy growing up in Connecticut in 1775 when his big brother Sam joins the Patriots and talks about becoming independent and free. But their father is a Loyalist, and Tim has to deal with a family split by war and politics. This book is both exciting and sad, and a very good read.

James Lincoln Collier & Christopher Collier, *War Comes to Willy Freeman; Jump Ship to Freedom; Who Is Carrie?* The Arabus Family Saga, Delacorte, 1983, 1981, 1984. Three excellent, related stories about some New England slave families caught up in the Revolutionary War. P.S. Willy Freeman is a girl.

Esther Forbes, *Johnny Tremain,* Houghton Mifflin, 1943. Smart, impatient Johnny's promising career as a silversmith is cut short by a horrible accident. He goes to work in a rebel-run print shop and is soon entangled in the events of the Revolution in Boston. This is a wonderful book, just as thrilling as it was when it was first published 50 years ago.

Jean Fritz, *Early Thunder,* Coward McCann, 1967. Daniel Webster is 14 in 1775. Salem, Massachusetts, where he lives, is fiercely divided between Patriots and Tory Loyalists. Daniel is a determined Tory, like his father, but the puzzling, exciting, scary events of the Revolution begin to make him think again.

Washington Irving, *Rip Van Winkle and The Legend of Sleepy Hollow,* Sleepy Hollow Press, 1980. Two stories from the Catskill Mountains. They were written in the 19th century, but are about even earlier times.

Scott O'Dell, *Sarah Bishop,* Houghton Mifflin, 1980. Sarah Bishop was a real English girl who came to the colonies shortly before the Revolution. Her family was killed in the battle of Long Island; Sarah ran away to the mainland and lived hidden in a cave in the woods. This book is well written and gives a very strong sense of the confusing mess that the war made of many people's lives.

Elizabeth Marie Pope, *The Sherwood Ring,* Houghton Mifflin, 1958. Peggy Grahame is a 20th-century teenager who goes to live with an eccentric uncle in her lonely family mansion in Orange County, New York. But the house is haunted by her thrilling colonial ancestors, who fought in the Continental army—and by their English opponents, too. This has a bit of mush and romance but it is mostly a great adventure story.

Ann Rinaldi, *Time Enough for Drums,* Holiday House, 1986. Sixteen-year-old Jem and her servant struggle to keep things going at home in Trenton, New Jersey, when the men of the family join the war of independence from the British king.

Elizabeth George Speare, *Calico Captive,* Houghton Mifflin, 1957. Based on a true story, this book tells of Miriam Willard, a girl who is captured with the Johnson family in an Indian raid in New Hampshire in 1754. They are marched through the wilderness to Canada and sold to the French in Montreal, where they are held for ransom.

Jean Fritz has written several fun-to-read biographies about important people in colonial times. All are highly recommended and published by Putnam.

And Then What Happened, Paul Revere?
Can't You Make Them Behave, King George?
What's the Big Idea, Ben Franklin?
Where Was Patrick Henry on the 29th of May?
Why Don't You Get a Horse, Sam Adams?
Will You Sign Here, John Hancock?

And also *The Great Little Madison* (1989). Don't miss this terrific biography of a very important American.

George Bernard Shaw, *The Devil's Disciple,* Penguin, 1950. In this play by a famous writer, a ne'er-do-well named Dick Dudgeon disguises himself as Patriot Parson Anderson, who is being hunted by redcoats, and finds that sacrificing himself for someone else is making him a better person. This is a very funny play and not hard to read.

Picture Credits

cover: Centennial gift of Watson Grant Cutter, Museum of Fine Arts, Boston, Massachusetts; p. 5: Musée de Versailles; p. 6: Independence National Historical Park, Philadelphia; pp. 6–7 (bottom): Library of Congress; p. 9: detail, Nathaniel Hurd, *Celebration of Britain's Heroes*, American Antiquarian Society; p. 10: Thomas Cole, *Autumn in the Catskills*, Arnot Art Museum, Elmira, New York; p. 11: Isaac Basire, *The Cherokee Embassy to England*, South Caroliniana Library, University of South Carolina; p. 12 (top left): Metropolitan Museum of Art, gift of the estate of James Hazen Hyde, 1959 (1974.201); p. 12 (bottom right): from *Davy's Crockett's Almanack*; pp. 13, 14: Library of Congress; p. 15: New-York Historical Society; p. 16: I. N. Phelps Stokes Collection, New York Public Library (# 1797-B-120); p. 19: Mount Vernon Ladies' Association of the Union; p. 20: courtesy John Carter Brown Library at Brown University; pp. 21–27: Library of Congress; p. 28: National Gallery of Canada; p. 29: Library of Congress; p. 31(top): William Hoare, *William Pitt*, Museum of Art, Carnegie Institute, Pittsburgh; p. 31 (bottom), engraving after Benjamin West, *The Death of General Wolfe*, Library of Congress; p, 32: Library of Congress; p. 34 (top): collection of Mr. Leavett-Shenley, The Holt, Upham, Engand; p. 34 (bottom): Sir Joshua Reynolds, *Sir Jeffery Amherst*, Mead Art Museum, Amherst College; p. 35: New York Public Library; p. 36 (top): Ñew York Public Library Picture Collection; p. 36 (bottom): Library of Congress; p. 38: unidentified artist, *The German Bleeds & Bears ye Furs*, Library Company of Philadelphia; p. 39: New York Public Library; p. 41: Library of Congress; p. 42: Charleston Museum; p. 43: Edward Greene Malbone, *Eliza Izard (Mrs. Thomas Pinckney, Jr.)*, Gibbes Museum of Art/Carolina Art Association; pp. 44–45: Charleston Museum; pp. 46–48: New York Public Library Picture Collection; p. 50: Library of Congress; p. 51 (top): Colonial Wiliamsburg Foundation; p. 51 (bottom): Library of Congress; p. 51 (inset): Board of Inland Revenue Library, London; p. 52 (top): Metropolitan Museum of Art, bequest of Charles Allen Munn, 1924 (24.90.1865); p. 52 (bottom): courtesy John Carter Brown Library at Brown University; p. 53: *The Repeal...*, Library Company of Philadelphia; pp. 54–57: Library of Congress; p. 60: Peter F. Rothermel, *Patrick Henry Before the Virginia House of Burgesses*, Red Hill–Patrick Henry National Memorial; p.61 (top right): courtesy John Carter Brown Library at Brown University; p. 62 (top): Historical Society of Pennsylvania; p. 62 (bottom): Monticello, Thomas Jefferson Memorial Foundation; p. 63 (left): Massachusetts Historical Society; p. 63 (right): American Antiquarian Society; p. 64: Mansell Collection; p. 64 (inset): Jeremiah Meyer, *General Thomas Gage*, National Portrait Gallery; p. 65 (top): Library Company of Philadelphia; p. 65 (bottom): *A Monumental Inscription to the Fifth of March*, Mansell Collection; p. 66: Library of Congress; p. 67 (top left): courtesy Winterthur Museum ; p. 67 (bottom left): Historical Society of Pennsylvania; p. 67 (right): Benjamin West, *George III as Commander in Chief of the Army*; reproduced by gracious permission of Her Majesty the Queen; p. 68: collection of General E. C. R. Lasher; p. 69 (bottom right): Library of Congress; p. 70 (bottom): Connecticut Historical Society, Hartford; p. 72: Prince Consort's Library, Aldershot, England; p. 74: Museum of Fine Arts, Boston; p. 75: Library of Congress; p. 76: New York Public Library Picture Collection; p. 78 (top left): Library of Congress; p. 78 (bottom right): Fort Ticonderoga Museum; p. 79 (top left): British Museum; p. 79 (bottom): National Portrait Gallery, London; p. 79 (right): Musée Jacquemart—André, Abbaye de Chaalis-Parnotte; p. 80: Mount Vernon Ladies' Association of the Union; p. 81: Library of Congress; p. 82: Independence National Historical Park, Philadelphia; p. 83 (middle): Maryland Historical Society; p. 84: New York Public Library Picture Collection; p. 85: Library of Congress; p. 86 (top): Library of Congress; p. 87 (top): American Antiquarian Society; p. 87 (bottom): New York Public Library Picture Collection; p. 88: Library of Congress; p. 89 (left): Library Company of Philadelphia; p. 89 (right): Library of Congress; p. 90 (left): Museum of Fine Arts, Boston; pp. 90–91: Mansell Collection; p. 92: Yale University Art Gallery, Trumbull Collection; p. 93 (right): American Philosophical Society; p. 94: British Museum; p. 95 (oval): Library of Congress; p. 95 (right): I. N. Phelps Stokes Collection (Stokes before 1739-B-63), New York Public Library; p. 96: Collection of the U. S. Senate; p. 97: Library of Congress; p. 98: Independence National Historical Park, Philadelphia; pp. 99, 100: Library of Congress; p. 101: collection of Henry Middleton Drinker; p. 102, p. 105: Library of Congress; pp. 106–107: Yale University Art Gallery, Garvan Collection; p. 108: Washington/Custis/Lee Collection, Washington and Lee University, Lexington, Virginia; p. 109 (top right): Massachusetts Historical Society; pp. 110–111: New York Public Library Picture Collection; p. 112: Yale University Art Gallery, Trumbull Collection; p. 113 (top): Library of Congress; p. 113 (bottom): Maryland Historical Society; p. 114 (bottom): courtesy Kennedy Galleries, Inc. , New York; p. 116: Library of Congress; p. 117: Musée de Versailles; p. 118 (left): Library Company of Philadelphia; p. 118 (right): Pennsylvania Academy of Fine Arts; p. 119: Library of Congress; p. 119 (inset): Connecticut Historical Society, Hartford; p. 120 (top): Colonial Williamsburg Foundation; p. 121 (top): Metropolitan Museum of Art, bequest of Charles Allen Munn, 1924 (24.90.1427); p. 121 (bottom right): Anne S. K. Brown Military Collection at Brown University; p. 122 (top right): Library of Congress; p. 123: Emmet Collection, New York Public Library (#7815); p. 126 (top): Maryland Historical Society; pp. 126 (bottom), 127: Library of Congress; p. 128 (top): *Military Antiquities—History of the English Army from the Conquest to 1801*; pp. 128 (bottom), 129: Library of Congress; p. 130 (top): (detail), Allen Memorial Art Museum, Oberlin College, R. T. Miller, Jr., Fund, 1946; p. 130 (bottom): Independence National Historical Park, Philadelphia; p. 131: New York Public Library Picture Collection; p. 133: Independence National Historical Park, Philadelphia; pp. 134 (bottom), 135: Historical Society of Pennsylvania; p. 136 (top): Independence National Historical Park; p. 136 (bottom): Bancroft Library, University of California, Berkeley; p. 138: New York Public Library Picture Collection; p. 140 (top): British Museum; p. 140 (bottom): National Portrait Gallery, London; p. 141: Library of Congress; p. 142 (left): Mariner's Museum, Newport News, Virginia; p. 142 (right): U.S. Naval Academy Museum; p. 143: Detroit Institute of Arts Founders Society, gift of Dexter M. Ferry, Jr.; p. 144 (top): collection of Harold L. Peterson; p. 145: Historical Society of Pennsylvania; pp. 147(bottom), 148 : Library of Congress; p. 150: Library of Congress; p. 152 (top): Yale University Art Gallery; p. 152 (bottom): Library of Congress; pp. 153, 154, 155 (top): Thomas Jefferson Memorial Foundation; p. 156 (bottom): Library of Congress; p. 157 (top): courtesy Albert E. Leeds; p. 157 (bottom): American Philosophical Society; p. 158: Library of Congress; p. 159: British Museum of Natural History; p. 160 (top): National Portrait Gallery; p. 160 (bottom): Free Library, Philadel-phia; p. 161: collection of Joseph and Sarah Harrison, Pennsylvania Academy of Fine Arts; p. 162: Maryland Historical Society; p. 163 (left): Library of Congress; p. 164 (bottom): Pennsylvania Academy of Fine Arts; p. 165 (top): Library Company of Philadelphia; p. 165 (bottom): Pennsylvania Academy of Fine Arts; p. 166 (top): New York Public Library Picture Collection; p. 166 (bottom): American Antiquarian Society; p. 167 (top): New-York Historical Society; p. 168: Historical Society of Pennsylvania; p. 169: Yale University Art Gallery, gift of Roger Sherman White; p. 170: Library of Congress; p. 171: courtesy John Carter Brown Library at Brown University; p. 172: Library Company of Philadel-phia; p. 174 (top): New York Public Library; p. 174 (bottom): New York Public Library Picture Collection; p. 175: Lewis Miller Sketch Books, Virginia State Library ; p. 176: Library of Congress; p. 177: Independence National Historical Park, Philadelphia; p. 178: American Philosophical Society; p. 179: Independence National Historical Park, Philadelphia; p. 180: Mead Art Museum, Amherst College; p. 182 (top): Independence National Historical Park, Philadelphia; p. 182 (bottom): Historical Society of Pennsylvania; p. 183: from Samuel Cooper, *History of North America*, New York Public Library; p. 184: Library of Congress.

George III, ex-king of America

Index

A Note from the Author

Exact imagining. That's what historians are supposed to do. I know that because I read it in a book by George Steiner. He said, "History is exact imagining." Now Steiner is an erudite (AIR-yoo-dite—it means brainy) English historian. But exact imagining? That sounded like an oxymoron to me (ox-ee-MOR-on—I love that word—it means an absurd contradiction, like a sad optimist).

And, even though I know Steiner is very learned, exact imagining seemed contradictory. Fiction is imagining. And history is supposed to be exact. But is there something imaginative about writing true stories about the past?

I thought about the way I write my history books. This is what I do: I pretend. In this book I pretended that I lived in the 18th century. I pretended that I was a Patriot. I pretended that I was a Loyalist. Whenever I could, I got help with the pretending. I traveled and observed. I visited Boston and Williamsburg and Monticello and Philadelphia and Charleston. I talked to people who know about the past; I asked questions. And I read books, lots and lots of books, to learn all the details I could. That way, when I pretended, I wouldn't be ignorant about it. I wanted my pretending to be as exact as possible.

Exact as possible! Maybe Professor Steiner does make sense. If you want to be a historian you have to *imagine* yourself into the time and place that you are writing about, and then you need to be as careful and well-informed and *precise* as you can be.

I found the same idea in a book by Francis Parkman (a historian from Boston who lived in the 19th century and wrote about the French and Indian War). Parkman said that facts—even if they are exact—can mislead you, unless you use your imagination to put yourself in the past and understand those facts in their time in history. For instance, now that we know the facts about how the Revolutionary War came out, most of us think the Loyalists were wrong. But if you zoom back in time to 1775, you will find many serious-thinking Americans who believe that Patriots are a bunch of hotheads and that breaking away from England, the greatest nation in the western world, is a very poor idea.

This is what Francis Parkman said in a book called *Pioneers of France in the New World*:

> *Faithfulness to the truth of history involves far more than a research...[into] facts. Such facts may be detailed with the most minute exactness, and yet the narrative, taken as a whole, may be unmeaning and untrue. The narrator must seek to imbue himself with the life and spirit of the time....He must himself be, as it were, a sharer or spectator of the action he describes.*

Now I've found that there is a problem with this. It is easy to imagine yourself a hero or heroine of the past, but that is not fair to history. You also have to try to understand people you don't like. Pretending to be a slave escaping to the British may be exciting, but if you want to understand and write about slavery, you also have to pretend to be a slave owner. You have to ask yourself why you own slaves and what will happen if you free them. And what are you going to do? And why?

Have you ever thought about writing history? You could start by investigating the history of your own family. Ask your parents or grandparents what life was like when they were children. After you finish asking that (and taking notes about it), then ask what they know of *their* grandparents' lives.

Most of us think that other people's families are interesting but ours isn't. But every family—rich, poor, or in between—has its own stories; and they are all worth telling. When did the people in your family come to America? Where did they come from? Why did they come? Can you find out details about their journey here? What about your neighbors—do you know their stories? Does your house have a story? Your school? Your town? Why don't you try some exact imagining?